TECHNOLOGY AND THE AMERICAN WAY OF WAR

T0262553

Thomas G. Mahnken

Technology and the American Way of War

Columbia University Press

New York

Columbia University Press
Publishers Since 1893
New York   Chichester, West Sussex
Copyright © 2008 Columbia University Press
Paperback edition, 2010
All rights reserved

Library of Congress Cataloging-in-Publication Data
Mahnken, Thomas G., 1965–
   Technology and the American way of war / Thomas G. Mahnken.
       p.   cm.
   Includes bibliographical references and index.
   ISBN 978–0–231–12336–5 (cloth : alk. paper) — ISBN 978–0–231–12337–2
   (pbk. : alk. paper) — ISBN 978–0–231–51788–1 (e-book)
   1. Military art and science—Technological innovations—United States. 2. United
States—Armed Forces—Technological innovations. 3. United States—History,
Military—20th century. 4. United States—History, Military—21st century. 5.
United States —Armed Forces—History—20th century. 6. United States—Armed
Forces—History—21st century. I. Title.

U43.U4M34 2008
355'.070973—dc22

2007050421

Columbia University Press books are printed on permanent and durable
   acid-free paper.
This book is printed on paper with recycled content.
Printed in the United States of America
c 10 9 8 7 6 5 4 3 2 1
p 10 9 8 7 6 5 4 3 2 1

References to Web sites (URLs) were accurate at the time of writing. Neither the
   author nor Columbia University Press is responsible for URLs that may have
   expired or changed since the manuscript was prepared.

To Deborah

Contents

Acknowledgments

This book grew out of a dialogue with James Warren of Columbia University Press in the late 1990s, a time with a temper far different from that of today. The September 11, 2001, terrorist attacks and the wars in Afghanistan and Iraq that followed led to a reconceptualization of the book and the addition of chapter 6. James got me on the road and saw me along much of the way; I am grateful to Peter Dimock for getting me to the finish line.

I began writing this book while a professor of strategy at the U.S. Naval War College in Newport, Rhode Island, and completed it while a visiting fellow at the Philip Merrill Center for Strategic Studies at Johns Hopkins University's Paul H. Nitze School of Advanced International Studies in Washington, D.C. Each is the finest institution of its kind, and each provided me support and inspiration without which I could not have finished this book. At the Naval War College, I am grateful to George Baer, John Maurer, Carol Keelty, and Cathy Hubert, as well as the students in my elective courses on innovation, network-centric warfare, and information technology. At SAIS, I am grateful to Eliot A. Cohen, Thomas A. Keaney, Thayer McKell, and Courtney Mata, as well as the students in my course "Understanding Military Technology."

I owe a special debt to James R. FitzSimonds of the Naval War College, a model colleague and collaborator. Jim has done more than anyone to shape my view of military technology. I had the pleasure of teaching a series of elective courses with him at the Naval War College. Over the past seven years we have also been joint investigators on a project exploring the attitudes of officers regarding the future of warfare.

On a personal level, the period since I began this book saw the birth of my son Thomas and of my daughter Rachel, and the death of my mother, Madeleine. It also saw me go places and do things I could scarcely have imagined, including two extended overseas deployments with the Navy. I wrote some parts of the manuscript in a hotel room in Bahrain, others in a tent in Kuwait.

I would like to thank Thomas Ehrhard, Thomas A. Keaney, Andrew W. Marshall, Barry D. Watts, and James R. FitzSimonds for reading portions of the manuscript. Their comments and suggestions have made this a better book. I am also grateful to my research assistants—Eleanore Douglas, Benjamin Klay, Ewan MacDougall, Ani Ahn, Greg Schnippel, and S. Rebecca Zimmerman—for making sense of my scattered and sometimes scatterbrained direction. Any errors of fact or interpretation are, of course, mine alone.

TECHNOLOGY AND THE AMERICAN WAY OF WAR

Introduction

HERE ARE TWO statements by two senior officers in armed forces of the United States, separated by more than three decades, that have much to tell us about technology and the role it plays in the American way of war. The first, by General William Westmoreland, dates to 1969 and the height of the Vietnam War. The second, by Admiral William Owens, dates to 2000, the dawn of a new millennium.

> On the battlefield of the future, enemy forces will be located, tracked, and targeted almost instantaneously through the use of data links, computer assisted intelligence evaluation and automated fire control. With first round probabilities approaching certainty, and with surveillance devices that can continually track the enemy, the need for large forces will be less important.[1]

> I believe the technology that is available to the U.S. military today and now in development can revolutionize the way we conduct military operations. That technology can give us the ability to see a "battlefield" as large as Iraq or Korea—an area 200 miles on a side—with unprecedented fidelity, comprehension, and timeliness; by night or day, in any kind of weather, all the time. In a future conflict, that means an Army

corps commander in his field headquarters will have instant access to a live, three-dimensional image of the entire battlefield.... The commander will know the precise location and activity of enemy units—even those attempting to cloak their movements by operating at night or in poor weather, or by hiding behind mountains or under trees. He will also have instant access to information about the U.S. military force and its movements, enabling him to direct nearly instantaneous air strikes, artillery fire, and infantry assaults, thwarting any attempt by the enemy to launch its own attack.[2]

Most obviously, the two statements illustrate the technological optimism that has historically animated U.S. defense planning. Westmoreland's words also remind us that recent discussions of transforming the U.S. military by emphasizing sensors and precision strikes are not novel, but rather represent a school of thought that goes back decades. The two quotations are also testimony to the fact that the realities of technology development and acquisition frequently belie optimistic predictions. Indeed, technological attempts to clear away the fog of war remain as elusive today as they did in the 1960s.

This book explores how technology interacted with the culture of the U.S. armed services in the six decades following World War II. It argues that although technology has in some cases shaped the services, particularly the development of nuclear weapons and long-range ballistic missiles, more often the services have molded technology to suit their purposes. The cultures of the services have in fact proven resilient in the face of technological challenges to their identity. The book also examines how the strategic environment—throughout much of the period, the competition with the Soviet Union—shaped both technology and organizational culture.

That environment, organizational culture, and technology interacted in often surprising ways. For example, the need for battlefield mobility in a war in Central Europe combined with the U.S. Army's cavalry culture of maneuver and the technology of the helicopter to produce airmobile divisions that played an important role in the Vietnam War. The threat posed by Soviet bombers to U.S. carrier battle groups, combined with the U.S. Navy's culture of distributed command and the rapid growth of information technology, spawned the concept of network-centric warfare. Similarly, the need to stop a Soviet armored assault in Central Europe spurred the development of precision weapons that matured only after the collapse of the enemy against which they were designed.

Strategic Culture and the American Way of War

The notion that there is a connection between a society and its style of warfare has a long and distinguished pedigree. In his history of the Peloponnesian War, Thucydides records that the Spartan king Archidamus and the Athenian *strategos*, or general, Pericles linked the capabilities of their military to the constitution of their states.[3] Writing at the beginning of the twentieth century, Julian Corbett drew a distinction between the German or "continental" and British or "maritime" schools of strategic thought, with the former focusing on war between land powers and the latter on a conflict between a sea power and a land power.[4] Basil H. Liddell Hart refined Corbett's argument, noting that Britain had historically followed a distinctive approach to war by avoiding large commitments on land and using sea power to bring economic pressure to bear against its adversaries.[5]

A nation's way of war flows from its geography and society and reflects its comparative advantage.[6] It represents an approach that a given state has found successful in the past. Although not immutable, it tends to evolve slowly. It is no coincidence, for example, that Britain has historically favored sea power and indirect strategies, nor that it has traditionally eschewed the maintenance of a large army. Israel's lack of geographic depth, small but educated population, and technological skill have produced a strategic culture that emphasizes strategic preemption, offensive operations, initiative, and—increasingly—advanced technology.[7] Australia's liminal geopolitical status, its continental rather than maritime identity, and its formative military experiences have shaped its way of war.[8]

The notion of a distinct American way of war is inextricably linked to Russell Weigley's book of the same name.[9] In it, Weigley argues that since the Civil War the U.S. armed forces have pursued a unique approach to combat, one favoring wars of annihilation through the lavish use of firepower. In his formulation, its main characteristics include aggressiveness at all levels of warfare, a quest for decisive battles, and a desire to employ maximum effort. By contrast, the American military has been uncomfortable waging war with constrained means for limited or ambiguous objectives.

Weigley's formulation is best seen as a statement of how the U.S. armed forces would *like* to fight wars. U.S. military experience is far more variegated than Weigley admits. As Brian M. Linn has noted, the U.S. armed forces have in fact favored strategies of attrition over annihilation. In addition, the United States has throughout its history pursued a much wider range of strategies than Weigley's formulation indicates, including deterrence and wars for limited aims.[10] Linn and others have noted that the U.S. military has a rich tradition of fighting small wars and insurgencies. Indeed, Max Boot went so far as to propose this tradition as an alternative American way of war.[11]

Strategic Culture

American strategic culture has several enduring features. The United States has displayed, for example, a strong and long-standing predilection for waging war for unlimited political objectives.[12] During the Civil War, President Abraham Lincoln and General Ulysses S. Grant fought to defeat utterly the Confederacy. During World War I, General John J. Pershing, the commander of the American Expeditionary Force, favored a policy of unconditional surrender toward Imperial Germany even as President Woodrow Wilson sought a negotiated end to the conflict.[13] In World War II, Franklin D. Roosevelt and his commanders were of one mind that the war must lead to the overthrow of the German, Japanese, and Italian governments that had started the war. In the current war against Al Qaeda and its supporters, there is no sentiment for anything approaching a negotiated settlement.

Just as Americans have preferred a fight to the finish, so, too, have they been uncomfortable with wars for limited political aims. In both the Korean and Vietnam wars, American military leaders were cool to the idea of fighting merely to restore or maintain the status quo. Indeed, General Douglas MacArthur likened anything short of total victory over communist forces on the Korean peninsula to "appeasement."[14] Similarly, the standard explanation of American failure in Vietnam—and the one most popular among U.S. military officers—is that the U.S. military would have won the war were it not for civilian interference.[15]

Related to the desire to wage war for unlimited political objectives is a tendency to demonize America's adversaries. Such a view is the product of U.S. history: during the twentieth century the United States fought a series of despotic regimes, from Hitler's Germany and Kim Il-Sung's North Korea to Saddam Hussein's Iraq and Slobodan Milosevic's Serbia. However, there is a clear tension between the need to rally the public in support of the use of force and the need to pursue limited aims. Political leaders who demonized America's adversaries often faced a backlash when the United States did not continue the war to the finish. Advisors to President George H. W. Bush, for example, bridled at his comparisons of Saddam Hussein to Adolf Hitler, fearing that it would complicate the conduct of the 1991 Gulf War.[16] And the United States has encountered difficulty when it has fought adversaries who at least appear less than demonic. Although Ho Chi Minh presided over a brutal communist government, North Vietnamese propaganda and American opponents of the war in Vietnam were able to portray him as a kindly "Uncle Ho," or even a latter-day George Washington. The United States is thus fortunate to have in the current war on terror an adversary such as Usama (or Osama) bin Laden, an individual who viscerally hates the United States and all it stands for and has made no effort to find favor among its people.

Reliance on advanced technology has been a central pillar of the American way of war, at least since World War II. No nation in recent history has placed greater emphasis upon the role of technology in planning and waging war than the United States. World War II witnessed the whole-sale mobilization of American science and technology, culminating in the detonation of the atomic bomb. Technology played an important role in America's conduct of the Cold War as well, as the United States sought to use its qualitative advantage to counterbalance the numerical superiority of the Soviet Union and its allies. America's post–Cold War conflicts in Iraq, the former Yugoslavia, and Afghanistan highlighted its technological edge over friend and foe alike.

Although the U.S. military as a whole favors technology, such a view has not gone unchallenged. To the contrary, civilian and military leaders and defense analysts have repeatedly debated the merits of the U.S. military's reliance on advanced technology. On one side have been technophiles who argued, explicitly or implicitly, that technology holds the key to victory in war. Arrayed against them are latter-day Luddites who decry the American military's seeming fascination with technology.

In the 1940s and 1950s, technophiles argued that the advent of nuclear weapons had revolutionized warfare. Opposing them were those who believed that nuclear weapons represented nothing more than very effective aerial bombardment or fire support—tools that made armies more effective at performing existing tasks rather than creating new roles and missions. In the 1970s and 1980s, the technophiles supported the development of a range of new technologies, including stealth and precision-guided munitions. Their opponents in the military reform movement argued for procuring larger numbers of less complex weapon systems. Recent debates over whether the information revolution portends a major change in the character and conduct of war are but the most recent manifestation of this debate.

The 2003 Iraq War and its aftermath provide ammunition for both sides of the debate. Advanced technology such as precision-guided munitions, unmanned aerial vehicles, and command and control systems contributed to the ability of U.S. and allied forces to overthrow Saddam Hussein's re-gime rapidly. On the other hand, technology has not provided the solution to the insurgency that has followed.

Each perspective has its flaws. The technophiles can be accused of ig-noring the nonmaterial dimensions of strategy. Technological proficiency is no substitute for strategic acuity. Indeed, technical prowess may breed hubris. The Luddites, however, can be accused of underplaying technol-ogy's benefits. For all the talk of how little technology matters, it is a rare soldier who would swap his state-of-the-art M1 *Abrams* main battle tank for even the best tank in the inventory of any plausible opponent.

As Colin Gray has observed, strategic culture is neither good nor bad. Rather, it represents the context for strategic action:

> The machine-mindedness that is so prominent in the dominant American "way of war" is inherently neither functional nor dysfunctional. When it inclines Americans to seek what amounts to a technological, rather than a political, peace, and when it is permitted to dictate tactics regardless of the political context, then on balance it is dysfunctional. Having said that, however, prudent and innovative exploitation of the technological dimen-sion to strategy and war can be a vital asset.[17]

America's traditional reliance upon technology in war is certainly no recipe for success. Technology is a poor substitute for strategic think-ing. The United States lost in Vietnam despite enjoying a considerable technological edge—at least in most areas—over its adversaries because it failed to develop an adequate strategy to achieve its political objectives. During the 1990s, the U.S. government increasingly looked to technology, in the form of air- and sea-launched precision-guided munitions, to solve problems—such as terrorism and ethnic violence—that were at their root political. Washington's penchant for advanced technology also fostered the illusion among some that the United States could use force without killing American soldiers and innocent civilians, and among America's enemies the impression that the United States was averse to sustaining casualties. Saddam Hussein, for one, saw high-technology warfare as a sign of American weakness rather than of American strength.[18]

Service Culture

Although American strategic culture has well defined features, each ser-vice also has its own unique culture, one shaped by its past and which, in turn, influences its current and future behavior.[19] Service cultures are hard to change, because they are the product of the acculturation of millions of service members over decades and are supported by a network of social and professional incentives. People join the U.S. Army, U.S. Navy, U.S. Air Force, and U.S. Marine Corps, not "the military" in the abstract. They do so because they identify—or want to identify—with a service's values and its culture. It is therefore not surprising that more than two decades after the passage of the Goldwater-Nichols Act, which sought to promote jointness, an officer's service affiliation remains the most important determinant of his or her views, more than rank, age, or combat experience.[20]

In many cases, service identity is more important to officers than branch identity. All aviators, for example, are not alike: Air Force pilots have cultural attitudes that differ significantly from those of their Navy counterparts.[21] Army infantry officers similarly have views that differ significantly from their Marine Corps counterparts.[22]

One way in which service culture manifests itself is in attitudes toward technology. For example, not all elements of the U.S. military are equally reliant on technology. Because war at sea and in the air is by definition technologically intensive, the Navy and Air Force have tended to emphasize the role of technology in war. The Army and Marines, by contrast, have tended to emphasize the human element. As the old saw goes, the Air Force and Navy talk about manning equipment, whereas the Army and Marine Corps talk about equipping the man. Not surprisingly, therefore, Army and Marine Corps officers tend to be more skeptical than their Air Force and Navy counterparts about the impact of technology on the character and conduct of war.[23]

The services also vary in terms of their structure and dominant groups. The Marines and Air Force are "monarchical," with powerful service chiefs drawn from a single dominant subgroup, whereas the Army and Navy are "feudal," with less powerful chiefs drawn from a variety of subgroups.[24] Each also has its own "altars of worship"—those things that the institution values.[25] These characteristics, in turn, affect how the services approach technology and how technology affects the service.

The U.S. Marine Corps is a unitary, monarchical organization. The smallest of the services, it is also the most cohesive. Its ethos is based on the notion that all marines are the same and that every marine is a rifleman. Despite the fact that the Marine Corps contains all combat arms—infantry, artillery, and armor—as well as an aviation component, all of the last ten Marine Corps commandants have been infantrymen.

Of the U.S. armed forces, the Marine Corps has the strongest commitment to tradition and the status quo, a commitment reinforced by the deliberate, self-conscious study of history. It is, for example, the only service that teaches history as part of Officer Candidate School.

The Marine Corps' emphasis on tradition and conformity is manifest in its uniform, which has changed the least since World War II compared to the uniform of any other service. It also reflects the service's ethic of conformity; with the exception of aviators, who wear gold flight wings on their chest, it is impossible to determine a marine's specialty from his or her uniform.

Marines value technology the least of any service. In part, this is the result of a culture that puts the individual warrior at the center of warfare. It is also the result of the fact that the Marine Corps has historically had the least money to devote to technology. Until very recently, the Marine Corps let the Army and Navy develop the majority of their equipment, adopting and adapting it as necessary. Not surprisingly, the Marine Corps figures the least prominently in the chapters that follow.

Power in the Army is shared among the traditional combat arms: infantry, cavalry/armor, and artillery. Not surprisingly, the position of chief of staff tends to rotate among these combat arms. The current one, General George Casey, is an infantry officer; his most recent ten predecessors included four from the infantry, three from the artillery, two from armor, and one from Special Forces.[26]

Whereas service identity is paramount to a marine, his or her counterpart in the Army attaches great importance to branch identity. The Army is, in Carl Builder's words, "A mutually supportive brotherhood of guilds. Both words, *brotherhood* and *guilds*, are significant here. The combat arms or branches of the Army are guilds—associations of craftsmen who take the greatest pride in their skills, as opposed to their possessions or positions. The guilds are joined in a brotherhood because, like brothers, they have a common family bond (the Army) and a recognition of their dependency upon each other in combat."[27] Unlike the Marine Corps, an Army officer's branch identity is visible on his or her uniform.

The Army has tended to assimilate technology into its existing branch structure. The widespread adoption of the helicopter, for example, did not spawn a new branch, but rather led to a redefinition of cavalry to include rotary-wing aircraft.

Army officers, like their marine counterparts, frequently profess that technology plays a subordinate role in warfare. In fact, however, the Army has traditionally valued advanced technology. Indeed, as the chapters that follow show, its leaders have consistently seen advanced technology as a comparative advantage over potential foes.

Technology is inherently more important to naval forces than to ground forces. Navies operate in an environment that is intrinsically hostile, and sailors from time immemorial have depended on naval technology to protect them from the elements. This has produced an attitude that recognizes the importance of technology but also prizes the tried-and-true over the novel.

The twentieth century witnessed the U.S. Navy's evolution from a monarchical to a feudal organization. At the dawn of the century, navies were synonymous with surface fleets. During the century, however, the development of naval aviation and submarine forces changed the structure of the Navy fundamentally. Whereas the Army has tended to assimilate new ways of war into existing branches, the Navy responded to the advent of aircraft and submarines by adding new branches and career paths. As a result, the dominant communities in the Navy are surface, submarine, and aviation. These three branches collectively control the Navy: of the last ten chiefs of naval operations, three have been aviators, four surface warfare officers, and three submariners.[28]

The Air Force had its origins in, and continues to be defined by, the technology of manned flight. The Air Force is divided into pilots and non-pilots and between different communities of pilots. Even though combat pilots make up less than one-fifth of the Air Force, they are the ones who have dominated the service since its inception.[29] From 1947 to 1982, the Air Force chief of staff was always a bomber pilot; since 1982, the chief of staff has always been a fighter pilot.

The Strategic Environment

Technology and service culture do not interact in a vacuum. Rather, the strategic environment provides the context that influences both. Throughout most of the period covered in this book, it was the challenge posed by the Soviet Union that drove the United States to develop and field new weapons. The need to deter a Soviet nuclear attack and defend Western Europe against the Warsaw Pact posed major challenges to which technology appeared to offer a solution. Civilian and military leaders throughout the Cold War viewed America's technological lead as a comparative advantage over the more numerous but less sophisticated arms of the Soviet Union.

Not surprisingly, the Cold War influenced the U.S. armed forces. It is difficult to imagine the United States developing forward-based, general-purpose armored and mechanized forces, let alone tens of thousands of nuclear weapons, more than a thousand intercontinental ballistic missiles, and highly advanced strategic reconnaissance aircraft and satellites absent the long-term competition with the Soviet Union.

The Cold War not only helped determine the size and shape of the U.S. armed forces, but it also influenced which parts of a service were more important than others. For the first half of the Cold War, the importance of strategic nuclear bombing meant that the Strategic Air Command (SAC) and its bomber generals dominated the U.S. Air Force; in the second half, the legacy of Vietnam and the challenge of fighting a war in Central Europe meant that it was the Tactical Air Command and its fighter pilots who reigned supreme. Similarly, the need to fight on NATO's Central Front led the U.S. Army to emphasize heavy armored and mechanized formations. The importance of nuclear attack and ballistic missile submarines and carrier aviation gave these communities prominence within the U.S. Navy.

The interaction between the United States and Soviet Union was far more complex than the action-reaction phenomenon international relations theorists posited. As Andrew W. Marshall wrote in 1972, "Commonly used hypotheses about the nature of the strategic arms race, or about the U.S.-Soviet interaction process (claiming a closely coupled joint evolution of U.S. and Soviet force postures), are either demonstrably false or highly suspect. The more serious classified studies of the interaction process almost uniformly present a picture of a much more complex, slower moving action-interaction process than that asserted by arms control advocates."[30]

The U.S.-Soviet competition provided a set of strategic and operational problems to guide the development of U.S. forces; it did not dictate the solution to them. Rather, it was the interaction of the environment with technology and service culture that shaped the services' choices.

About This Book

This book is about the interaction of technology and culture in the context of the strategic environment. It argues that technology both shaped and was shaped by the culture of the U.S. armed services. On the one hand, technology undoubtedly shaped the U.S. military. Most dramatically, the advent of nuclear weapons and long-range ballistic missiles changed—in some cases dramatically—the structure and organization of the armed services. It led, for example, to the formation of the Strategic Air Command and the development of ballistic missiles by the Army, Navy, and Air Force.

On the other hand, the culture of the U.S. armed services influenced the technologies that they chose to pursue. Technology does not dictate solutions. Rather, it provides a menu of options from which militaries choose. A service's culture, in turn, helps determine which options are more or less attractive.[31] During the early Cold War, the Air Force's preference for manned aircraft over cruise missiles, and for cruise missiles over ballistic missiles, affected how it went about exploiting the nuclear revolution. A service's culture also shapes how its officers view new technologies. Air Force officers, for example, viewed missiles as unmanned bombers, whereas Army officers saw them as long-range artillery.

On balance, the services shaped technology far more than technology shaped the services. Indeed, the culture of the services proved to be resilient in the face of technological threats. Even such a disruptive development as the advent of nuclear weapons left the culture of the services generally intact. The Army undertook a radical change in the late 1950s in fielding the Pentomic Division, a formation optimized for the nuclear battlefield, only to scrap it beginning in 1960 and return to an organizational structure reminiscent of that it had in World War II. The Navy's carrier battle groups survived challenges from land-based bombers, seaplanes, and nuclear ballistic missile carrying submarines. Indeed, the resilience of the services in the face of such dramatic changes offers a cautionary tale for those who seek to transform service culture.

There are, however, limits to the thesis that the culture of the armed services shaped technology choices. One is, for example, at a loss to find an example in the last six decades of American military history of a weapon system that was adopted or rejected purely due to service preferences. Often technological feasibility and service preferences go hand in hand. If there are narrow, parochial reasons for opposing new systems, there are also often sound technical reasons for doing so.

The book is framed by two military revolutions: the nuclear revolution of the late 1940s and 1950s and the ongoing information revolution.[32] It is also punctuated by a series of wars: not only the Cold War, but also hot wars in Korea, Vietnam, the Persian Gulf, the former Yugoslavia, Afghanistan, and Iraq.

Chapter 1 discusses the nuclear revolution, which spanned the decade and a half following the end of World War II. It explores how the competition with the Soviet Union shaped the U.S. armed forces during the early Cold War. It also examines how each of the services responded to the advent of nuclear weapons and long-range delivery vehicles such as bombers and missiles.

Chapter 2 explores the evolution of the U.S. armed forces between 1961 and 1975, a period dominated by the strategy of flexible response and the rise of civilian control over the Department of Defense. It discusses how the services adapted to flexible response. It also examines U.S. nuclear modernization and the development of ballistic missile defenses. In each case, interaction among the services and between the services and the office of the secretary of defense played a central role.

Chapter 3 examines America's use of advanced technology during the Vietnam War. It also discusses the services' use of innovative technology during the conflict, including the Army's airmobile units, the Navy's riverine force, and the Air Force's gunships. It also describes the introduction of a range of new technologies, including unmanned aerial vehicles (UAVs), unattended ground sensors, and precision-guided munitions (PGMs), that would prove their worth in the wars of the 1990s and beyond.

Chapter 4 discusses the role of technology in the late Cold War, from 1975 to 1990. It explores the debate over the role of technology in U.S. strategy brought on by the "military reform" movement. It examines the development of new weapons by the Army, Navy, and Air Force as well as the modernization of U.S. strategic and intermediate-range nuclear forces (INF) and the advent of the Strategic Defense Initiative (SDI).

Chapter 5 discusses the use of these weapons on the battlefield between 1991 and 2001, after the Cold War had ended. The demonstrated effectiveness of PGMs and stealth 1991 Gulf War and the conflicts that followed led many to argue for the emergence of a revolution in military affairs and the emergence of a "new American way of war."

Chapter 6 examines the reemergence of the traditional American way of war, with its use of massive force to overthrow the nation's foes, albeit with means far different from those available to previous generations. It examines the role of technology in the wars in Afghanistan and Iraq. Chapter 7 attempts to glean insights from the past six decades for the contemporary debate over the prospect of a revolution in military affairs brought on by the information revolution, as well as the role of the U.S. armed forces in exploiting that revolution.

Notes

1. Quoted in Paul Dickson, *The Electronic Battlefield* (Bloomington: Indiana University Press, 1976), 71.
2. Bill Owens and Ed Offley, *Lifting the Fog of War* (New York: Farrar, Straus & Giroux, 2000), 14–15.
3. Robert B. Strassler, ed., *The Landmark Thucydides: A Comprehensive Guide to the Peloponnesian War* (New York: Free Press, 1996), 45–46, 81–82.

4. Julian S. Corbett, *Some Principles of Maritime Strategy* (London: Longmans, Green, 1911), 38.

5. Basil H. Liddell Hart, *The British Way in Warfare* (New York: Macmillan, 1933).

6. Colin S. Gray, *Modern Strategy* (Oxford: Oxford University Press, 1999), chapter 5.

7. Michael I. Handel, "The Evolution of Israeli Strategy: The Psychology of Insecurity and the Quest for Absolute Security," in *The Making of Strategy: Rulers, States, and War*, ed. Williamson Murray, MacGregor Knox, and Alvin Bernstein, 534–78 (Cambridge: Cambridge University Press, 1994).

8. Michael Evans, *The Tyranny of Dissonance: Australia's Strategic Culture and Way of War, 1901–2005* (Canberra: Land Warfare Studies Center, 2005).

9. Russell F. Weigley, *The American Way of War: A History of United States Military Strategy and Policy* (Bloomington: Indiana University Press, 1973). For a more critical appraisal of the "American way of war," see Colin S. Gray, "Strategy in the Nuclear Age: The United States, 1945–1991," in Murray, Knox, and Bernstein, *The Making of Strategy*, 579–613.

10. Brian M. Linn, "*The American Way of War* Revisited," *Journal of Military History* 66, no. 2 (April 2002): 501–33.

11. Max Boot, *The Savage Wars of Peace: Small Wars and the Rise of American Power* (New York: Basic Books, 2002).

12. As Clausewitz wrote, "War can be of two kinds, in the sense that either the objective is to *overthrow the enemy*—to render him politically helpless or militarily impotent, thus forcing him to sign whatever peace we please; or *merely to occupy some of his frontier districts* so that we can annex them or use them for bargaining at the peace negotiations. Transitions from one type to the other will of course recur in my treatment; but the fact that the aims of the two types are quite different must be clear at all times, and their points of irreconcilability brought out" (emphasis in original). Carl von Clausewitz, *On War*, ed. and trans. Michael Howard and Peter Paret (Princeton, NJ: Princeton University Press, 1976), 69.

13. David R. Woodward, *Trial by Friendship: Anglo-American Relations, 1917–1918* (Lexington: University Press of Kentucky, 1993), 213–14.

14. See the testimony of General Douglas MacArthur in Allen Guttmann, ed., *Korea: Cold War and Limited War*, 2nd ed. (New York: D. C. Heath, 1972).

15. For alternative views, see Eliot A. Cohen, *Supreme Command: Soldiers, Statesmen, and Leadership in Wartime* (New York: Free Press, 2002), 175–84; Andrew Krepinevich, *The Army and Vietnam* (Baltimore: Johns Hopkins University Press, 1986).

16. George Bush and Brent Scowcroft, *A World Transformed* (New York: Knopf, 1998), 389.

17. Gray, *Modern Strategy*, 147.

18. Kevin M. Woods, Michael R. Pease, Mark E. Stout, Williamson Murray, and James G. Lacey, *Iraqi Perspectives Project: A View of Operation Iraqi Freedom from Saddam's Senior Leadership* (Norfolk, VA: U.S. Joint Forces Command, 2006), 15.

19. Carl H. Builder, *The Masks of War: American Military Styles in Strategy and Analysis* (Baltimore: Johns Hopkins University Press, 1989), 7.

20. Thomas G. Mahnken and James R. FitzSimonds, *The Limits of Transformation: Officer Attitudes Toward the Revolution in Military Affairs* (Newport, RI: U.S. Naval War College, 2003), 108.

21. For example, when surveyed in 2002, 41 percent of Air Force pilots but only 21 percent of Navy aviators agreed with the statement, "The ability to strike an adver-

sary with precision weapons from a distance will diminish the need for the U.S. to field ground forces."

22. For example, when surveyed in 2002, 57 percent of Army infantry officers but only 30 percent of Marine infantry officers agreed with the statement, "The U.S. armed forces must radically change their approach to warfare to compete effectively with future adversaries." Sixty-five percent of Army infantry officers but only 14 percent of Marine infantry officers agreed with the statement, "Modern conditions require significant changes to traditional Service roles and missions."

23. Mahnken and FitzSimonds, *The Limits of Transformation*, 60.

24. Thomas P. Ehrhard, "Unmanned Aerial Vehicles in the United States Armed Services: A Comparative Study of Weapon System Innovation" (Ph.D. dissertation, Johns Hopkins University, 2000), 75.

25. Builder, *Masks of War*, 18.

26. Generals Fred C. Weyand, Bernard W. Rogers, Edward C. Meyer, and John A. Wickham Jr. were infantrymen; Gordon R. Sullivan and Eric K. Shinseki were tankers; William C. Westmoreland, Carl E. Vuono, and Dennis J. Reimer were artillerymen; and Peter Schoomaker was from the Special Forces.

27. Builder, *Masks of War*, 33.

28. Admirals James L. Holloway III Thomas B. Hayward, and Jay L. Johnson were aviators; Jeremy R. Boorda, Vern Clark, and Michael Mullen were surface warriors; and James D. Watkins, Carlisle A. H. Trost, and Frank B. Kelso II were submariners.

29. Ehrhard, "Unmanned Aerial Vehicles," 89.

30. A. W. Marshall, *Long-Term Competition with the Soviets: A Framework for Strategic Analysis*, R-862-PR (Santa Monica, CA: Rand Corporation, 1972), vi.

31. Thomas G. Mahnken, "Beyond Blitzkrieg: Allied Responses to Combined-Arms Armored Warfare During World War II," in *The Diffusion of Military Technology and Ideas*, ed. Emily O. Goldman and Leslie C. Eliason, 246–50 (Stanford, CA: Stanford University Press, 2003).

32. On the topic of military revolutions, see Eliot A. Cohen, "A Revolution in Warfare," *Foreign Affairs* 75, no. 2 (March–April 1996): 37–54; Andrew F. Krepinevich, "Cavalry to Computer: The Patterns of Military Revolutions," *The National Interest* (Fall 1994): 30–42; and Williamson Murray, "Thinking About Revolutions in Military Affairs," *Joint Force Quarterly* 12 (Summer 1997): 69–76.

The Nuclear Revolution, 1945–1960

THE END OF World War II and the onset of the Cold War presented American soldiers and statesmen with a series of challenges. Perhaps the greatest was the need to craft a strategy and develop forces in response to a new security environment characterized by competition with the Soviet Union. At the same time, the U.S. military confronted the imperative of adapting to the advent of nuclear weapons. These dual challenges drove the U.S. armed forces to implement sweeping changes in the period from 1945 to 1960.

Nuclear weapons were Janus-faced, offering both opportunity and challenge. On one hand, the United States possessed first a monopoly, then a commanding lead, in nuclear weapons. The atomic bomb seemingly offered Washington the ability to counterbalance Moscow's large conventional ground forces. On the other hand, the Soviet Union's acquisition of nuclear weapons in 1949 and of long-range delivery means soon thereafter rendered the United States vulnerable to attack. The nuclear competition with the Soviet Union pushed the United States to invest in a vast air and space reconnaissance effort. Even more dramatically, the need to take advantage of the nuclear revolution, while also dealing with the vulnerability it created, led to a wholesale change in the way the United States organized, trained, and equipped its armed forces.

The nuclear revolution challenged the identity of each of the services. Should the Air Force be organized around manned bombers, or missiles? Would the ballistic-missile submarine supplant the aircraft carrier as the central component of the Navy? It also called into question the utility of traditional land campaigns and large-scale amphibious operations and—by extension—the very existence of the Army and Marine Corps. Nuclear weapons offered the armed services the opportunity to develop new capabilities and triggered a vigorous competition over new missions, including long-range missiles, space reconnaissance, and continental air defense. The nuclear age witnessed the rise of new elites within the services, such as missile operators in the Air Force and nuclear submariners within the Navy. By embracing the nuclear attack mission, the Air Force garnered the lion's share of defense resources. By contrast, the Navy and, particularly, the Army were forced to justify their existence in nuclear terms.

Of course, not all innovations proved successful. Some, such as the Air Force's development of a nuclear-powered aircraft and the Army's pursuit of the pentomic division, led to dead ends. Other innovations—such as the strategic cruise missile—were failures in the 1950s but successes in later years.

In the end, the services proved resilient in the face of the nuclear challenge to their identity, structure and mission. Not only did technology influence the culture of the services, but the ethos of a service also shaped the technologies it pursued. For example, the Air Force and Navy preferred manned aircraft to missiles. The Army, for its part, tended to view nuclear weapons as highly effective artillery. Thus, although each of the services changed markedly in the decade and a half after World War II, in the end their prenuclear identities proved more resilient than technological enthusiasts would have predicted.

The Birth of the U.S.–Soviet Competition

The U.S. military entered the Cold War with a sense of confidence. Many defense experts felt that the United States would enjoy a monopoly on atomic weaponry for some time to come. General Leslie Groves, the former head of the Manhattan Project, which had produced the atomic bomb, believed that the Soviets would be unable to break the American atomic monopoly for two decades.[1] Vannevar Bush, the former chairman of the Office of Scientific Research and Development under Franklin D. Roosevelt, agreed that the American nuclear monopoly was durable. As he wrote in 1949, "To build a large stock of atomic bombs is an undertaking that will strain the resources of any highly industrialized nation."[2] The general belief among experts was that the threat of nuclear devastation would be so effective as to deter virtually all military challenges.

In fact, the American monopoly would last less than five years. On September 3, 1949, an Air Force WB-29 aircraft, flying between Japan and Alaska as part of the Air Force's nuclear reconnaissance program, detected an unusual amount of radioactive particles in the atmosphere.[3] This debris came from the Soviet Union's first atomic test, which had occurred on August 29. Intelligence analysts rapidly understood the significance of the discovery.

Technology for Strategic Reconnaissance

The birth of the U.S.-Soviet nuclear competition, the secretive nature of the Soviet regime, the geographic depth of the Soviet Union, and the U.S. technological base all spurred the United States to pursue air and space reconnaissance technology. Reconnaissance aircraft and satellites helped the United States determine the size and composition of the Soviet nuclear arsenal and would also warn of impending attack.

The U.S. air and space reconnaissance program pushed the realm of technological possibility. It resulted in aircraft able to fly higher and faster than previous models. It spawned the ability not only to put a satellite in orbit, but also to photograph specific points on the ground and return the film safely to Earth. It also laid the groundwork for a technical collection infrastructure that continues to provide the bulk of U.S. intelligence more than a decade and a half after the end of the Cold War and the collapse of the Soviet Union.

The U.S. strategic reconnaissance program was a response to the U.S. government's lack of information on the size and characteristics of the Soviet atomic program during the early years of the Cold War. Most intelligence on Soviet industry came from captured German aerial photographs and old maps. Such sources were of limited value in uncovering evidence of Soviet atomic research and development. Somewhat more helpful were debriefings of German scientists whom the Soviets captured at the end of World War II and later repatriated.[4] Despite such tidbits, the United States entered its nuclear competition with the Soviet Union essentially blind.

Understanding Soviet nuclear research and development required the ability to overfly the Soviet Union. Early efforts included the Air Force's Genetrix (WS-119L) program, which beginning in 1956 sent high-altitude balloons equipped with cameras drifting across Soviet territory. The program yielded meager results, however: the Air Force launched 516 balloons but was able to recover only 44 of them. Moreover, the U.S. violations of Soviet airspace sparked strong Soviet protests.[5]

Manned aircraft, and then unmanned satellites, would yield more useful information on Soviet military developments. Beginning in 1948, RB-29 reconnaissance aircraft assigned to the Air Force's 72nd Strategic Reconnaissance Squadron carried out a series of photographic and electronic reconnaissance missions over the Soviet Arctic and Far East. Although their cameras allowed them to photograph Soviet territory from international airspace, flights along the Soviet periphery revealed little. With presidential approval, in early August the squadron began flying over Soviet territory. The first such mission, on August 5, lasted more than nineteen hours. Flying at altitudes of 35,000 feet or more, such aircraft used onboard instruments to identify and exploit gaps in the Soviet radar coverage and penetrate Soviet airspace. Although the Soviets sometimes detected the missions and scrambled fighters to intercept them, it was not until 1949 with the appearance of the MiG-15 Fagot that Moscow possessed an aircraft capable of intercepting the RB-29.[6]

During the early 1950s the Air Force used the RB-50 (a modification of the RB-29), the RB-45 Tornado, and later the RB-47 Stratojet to fly electronic intelligence (ELINT) missions over the Soviet Union. During these missions, aircraft would identify potential targets through their electronic emissions. They would also map Soviet radar coverage, giving SAC bombers the information they would need to penetrate Soviet air space in the event of nuclear war. Successful missions depended upon locating gaps in the Soviet radar network by monitoring Soviet radio and radar traffic. While such flights yielded valuable intelligence, they were extremely dangerous: between 1950 and 1959, the Air Force and Navy lost at least sixteen aircraft, with 164 crewmen killed on such missions.[7]

In order to reduce the risk to U.S. aircrews of such operations, the United States pursued space-based ELINT collection. Indeed, the first U.S. intelligence satellite was an ELINT satellite. The satellite, launched under the cover of the Galactic Radiation and Background (GRAB) project, was ostensibly designed to measure solar radiation. Its real purpose, known to fewer than two hundred people, was gathering signals from Soviet air defense radars that Navy and Air Force ELINT aircraft could not observe. The satellite, which orbited five hundred miles above the earth between July 1960 and August 1962, received radar signals and transmitted them to collection sites on the ground. Analysts on the ground recorded these transmissions and flew them to the Naval Research Laboratory in Washington, D.C., where they were evaluated before being distributed to the National Security Agency.[8] GRAB was followed by the Poppy satellite, which collected radar emissions from Soviet naval vessels from December 1962 to August 1977.[9]

The Bomber Gap and Strategic Air Reconnaissance

The need to determine the extent of the Soviet nuclear arsenal, as well as the size of the bomber force that would deliver it, drove the United States to develop innovative reconnaissance aircraft, such as the U-2 and A-12 Blackbird. The images these aircraft took helped deflate estimates of the Soviet bomber fleet.

The limited quantity and suspect quality of information on the Soviet nuclear buildup in the early 1950s led the U.S. intelligence community to overestimate the size of the Soviet bomber force. During the 1954 May Day parade the Soviet government for the first time revealed the existence of the M-4 Bison strategic bomber.[10] That same year the jet-powered Tu-16 Badger appeared. The bomber's emergence only a year after that of the first American jet bomber, the B-47, shocked U.S. intelligence analysts. A year later the Tu-95 Bear made its debut at the Tushino Air Show.

Western observers monitoring rehearsals for the 1955 Soviet Aviation Day parade reported seeing between twelve and twenty Bisons. In fact, the Soviets flew the same ten aircraft around the viewing stand in various formations in order to give the impression that at least twenty bombers were participating in the fly-by. Taking the bait, the U.S. intelligence community revised its estimate of Soviet bomber production sharply upward, concluding that the Soviets had at least thirty strategic bombers in their inventory. In fact, only ten planes in the class were in flying order. The Defense Department pegged the USSR's monthly bomber production at six, rising to twenty by late 1956.[11] In fact, between January 1955 and June 1956 the Soviets manufactured a grand total of thirty-one of the aircraft. Between 1955 and 1958 the Bison suffered nine major accidents, leading the Soviets to withdraw it from service for a time.[12]

Photographs taken by the U-2 reconnaissance aircraft provided much of the evidence that deflated U.S. intelligence estimates of the Soviet bomber force. In March 1954 Lockheed sent Air Force Brigadier General Bernard Schriever an unsolicited proposal, CL-282, to produce a single-engine reconnaissance aircraft capable of flying at 70,000 feet with a 2,000-mile range. On November 27, 1954, President Eisenhower authorized $35 million for Project Aquatone under the direction of Richard Bissell of the CIA

and Air Force Colonel Osmond Ritland, with Lockheed as the prime contractor.[13] The U-2 was truly a team effort: Lockheed designed and built
the aircraft and provided ground crews; the Air Force recruited pilots,
planned missions, and ran operations; and the CIA oversaw the production of the plane and its cameras, chose and protected bases, and processed film.[14] The development of the aircraft was a closely held secret,
with only a handful of people knowing of its existence. This secrecy gave
Bissell a considerable amount of freedom; he used "unvouchered funds" to
simplify the competitive bidding process and speed up procurement.[15]

The U-2 made its first flight on August 5, 1955, less than ten months
after it had been authorized. On its first operational mission on June
20, 1956, a U-2 flew over East Germany and Poland. Between June 1956
and May 1960, it overflew the Soviet Union twenty-four times. U-2 photographs of Saratov-Engels airfield at Ramenskoye, southeast of Moscow, taken on July 5, 1956, put to rest the bomber gap. The images of the
only Bison base showed fewer than three dozen of the bombers when
the Air Force estimated that the Soviet Union possessed nearly a hundred.[16] As a result of the information the U-2 collected, by the spring of
1957 estimates of the size of the Soviet bomber force began to decrease.
By November 1959 the intelligence community projection of the Soviet
bomber force for mid-1961 was less than 20 percent of what it had been
in 1956.[17]

Although the U-2 was a valuable intelligence collection platform, its
designers knew that it would not be able to fly over the Soviet Union
with impunity forever. All they hoped for was a few good years. To pro-
long the aircraft's effective life by reducing the aircraft's radar cross-section and improving its survivability, Lockheed launched Project Rain-
bow.[18] The manufacturer introduced a number of modifications to the
U-2, including the use of iron ferrite paints to absorb radar beams and
the development of "black boxes," or electronic countermeasures (ECM),
for the aircraft.[19]

The successor to the U-2 was Project Oxcart, which produced the supersonic A-12 Blackbird. Whereas the U-2 derived its survivability from
its ability to fly high, the A-12 used speed to survive. Designing an aircraft to go several times the speed of sound over long distances forced
designers to literally go back to the drawing board. Because a standard
aluminum airframe would melt at the temperatures the aircraft would experience, designers needed to build the plane out of titanium. In addition,
the aircraft needed lots of fuel. Indeed, it was essentially a flying fuel tank.
These requirements inevitably took their toll, and by the end of 1960 the
project was 30 percent over budget and at least a year behind schedule.[20]

The Blackbird's first successful test fight came in April 1962. The aircraft featured a number of innovations, including the first deployed astronavigation system: the aircraft used a small computer-driven telescope that looked through a small window in the rear of the airplane to navigate according to a database of some sixty stars.[21] It was also one of the first aircraft designed to minimize its radar cross-section (RCS).[22]

The Missile Gap and Space Reconnaissance

Despite the success of the U-2 and A-12/SR-71, the days of manned reconnaissance over the Soviet Union were numbered. The crisis that resulted from the downing of Francis Gary Powers's U-2 in May 1960 both led to an end to overflights of the Soviet Union (though not of other countries) and gave impetus to space reconnaissance. Reconnaissance from space would, in turn, give the United States a much more accurate picture of the Soviet intercontinental ballistic missile (ICBM) program.

The United States lacked reliable information on the Soviet missile program in the early 1950s. One March 1953 CIA report on the Soviet bloc assessed the agency's understanding of the Soviet nuclear stockpile as "reasonably adequate." However, the authors admitted that "knowledge of current Soviet guided missile programs is poor, although certain projects based on German developments are fairly well known."[23] The following year, the National Intelligence Estimate (NIE) on Soviet strategic forces, NIE 11–6–54, admitted that the United States had no firm current intelligence on what guided missiles the Soviets were developing. U.S. assessments were based upon the intelligence community's knowledge of German missile programs, U.S. programs, and estimated Soviet capabilities in related fields. "Therefore our estimates of missile characteristics and dates of missile availability must be considered as only tentative, and as representing our best assessment in the light of inadequate evidence and in a new and largely unexplored field."[24]

Uncertainty over the size of the Soviet bomber and missile programs, combined with the legacy of Pearl Harbor, spurred fear of surprise attack. In February 1957 the Technological Capabilities Panel of the Office of Defense Mobilization Science Advisory Committee, chaired by James R. Killian Jr., the president of the Massachusetts Institute of Technology, presented President Eisenhower with its report, "Meeting the Threat of Surprise Attack." The report warned of the Soviet Union's growing capacity to launch a surprise attack and strongly recommended that the United States take steps to protect its strike force through better intelligence, early warning, and defensive measures. It also called on the administration to begin or accelerate the development of intermediate-range ballistic missiles (IRBMs) and ICBMs.[25]

Moscow's launch of the world's first artificial satellite on October 4, 1957, served as a wakeup call. If the Soviets could put a satellite into orbit, they could also hurl a nuclear warhead towards the United States. In November 1957 the Gaither Committee Report concluded that by 1959 the Soviet Union might have the ability to launch an ICBM attack against the United States. The group, which consisted of leading scientists, businessmen, and military experts under the leadership of H. Rowan Gaither Jr., the president of the Rand Corporation, warned that Strategic Air Command (SAC) bomber bases were highly vulnerable and recommended improvement to U.S. early warning systems, acceleration of ballistic missile and active and passive defense programs. In fact, concern over the Soviet bomber threat had already led SAC to institute procedures to allow its bombers to take off quickly on warning of attack. However, such measures were inadequate. The highly secret Sprague report, prepared in conjunction with the Gaither Committee, found that on a randomly selected day not a single plane could have gotten off the ground within six hours of warning. SAC could probably have gotten bombers carrying only 50 to 150 nuclear weapons in the air, and Soviet air defenses would likely have destroyed many of them.[26]

Soviet advances in space and ballistic missiles led many prominent officers to question the durability of American technological superiority. In 1959, Army General Maxwell D. Taylor wrote that the United States had lost its military edge over the Soviet Union in many fields. As he put it, "My personal conclusion is that until about 1964 the United States is likely to be at a significant disadvantage against the Russians in terms of numbers and effectiveness of long-range missiles—*unless heroic measures are taken now.*"[27] Army Lieutenant General James Gavin went even farther, stating that the Soviet Union had established a "clear advantage" over the United States technologically.[28] He saw an opportunity to use American technological resources to better advantage than had been the case up to that point. As he argued, "It would appear to be entirely possible for a nation to develop and carry out a well-conceived strategic plan in technology that could cause an opponent to waste vast amounts of critical resources."[29] Such a strategy "offers us a prospect of recovering the technological initiative and this we must do."[30]

The United States enlisted technology in the service of understanding Soviet missile developments, establishing a chain of radar stations along the Soviet border to monitor the Soviet missile flight test. By the end of the 1950s these ranged from Europe through Turkey to Iran and Pakistan.[31] In addition, in 1958 the United States and Norway began operating a station, code-named Metro, to intercept telemetry from Soviet missile tests.[32]

By 1960, communications and human intelligence led U.S. intelligence analysts to suspect that the Soviets were constructing an ICBM launch site at Plesetsk. On May 1, 1960, Francis Gary Powers's U-2 was sent to investigate the site. It was on this ill-fated mission that his aircraft was shot down by a V-750 (SA-2 Guideline) surface-to-air missile.[33] If his flight had not ended in tragedy, it likely would have produced photos of what was then the only operational ICBM launch facility in the Soviet Union, photos that would have revealed that Soviet deployments were not as advanced as the U.S. intelligence community believed.

The demonstrated vulnerability of the U-2 gave an impetus to the development of reconnaissance satellites. Unlike manned reconnaissance aircraft, whose antecedents stretched back to World War I, unmanned satellites were truly novel. The task of photographing the earth's surface from hundreds of miles away while orbiting at hundreds of miles per hour, however, was extremely challenging. That the United States was able to overcome such obstacles was a testament both to the scientific and technical ingenuity of the United States as well as the ability of the U.S. government to harness that ingenuity in the service of national goals.

The first operational photoreconnaissance satellite, code-named Corona, had its origin in a technologically ambitious proposal developed by the Rand Corporation for potential military uses of space. The result was a comprehensive Air Force space reconnaissance effort known as Weapon System 117L (WS-117L) or the Satellite and Missile Observation Satellite (SAMOS).[34] The program used the same streamlined management approach as the U-2, with close cooperation among the CIA, Air Force, and defense contractors such as Lockheed, Itek, General Electric, Kodak, and Douglas Aircraft.[35]

Between 1960 and 1972 the United States launched four versions of Corona, designated Keyhole (KH) 1 through 4.[36] The first successful Corona mission, launched on August 18, 1960, took photos of Mys Schmidta airfield in the Soviet Union. Its film capsule was snatched out of midair by a modified C-119 Flying Boxcar aircraft.[37] In the years that followed, 120 of 145 Corona missions were complete or partial successes. The satellites exposed over 2.1 million feet of film, took more than 800,000 pictures, and photographed a land area of 557 million square miles.[38]

Reconnaissance satellites gave the United States the ability to observe Soviet territory without violating the USSR's airspace and thereby putting American lives at risk. The intelligence they produced gave the U.S. government considerable insight into the Soviet ICBM program and greatly lessened concern over a Soviet surprise attack. Corona confirmed the existence of an ICBM base at Plesetsk, but also showed that many sites suspected of ICBM activity were in fact innocent. Corona was not, however, the only source of information on the Soviet missile program. The United States also benefited greatly from information provided by Lieutenant Colonel Oleg Penkovsky, who spied for the British Secret Intelligence Service. In the spring of 1961, Penkovsky provided information revealing that the Soviets had exaggerated the size of their ICBM program.[39] By September 1961, in the first NIE to incorporate intelligence gathered by Corona, the intelligence community had scaled back its estimate of SS-6s deployed to just over ten. Even this lower figure turned out to be an overestimate, as the Soviets had only deployed four of the unwieldy mammoths.[40]

Nuclear Weapons and the U.S. Armed Forces

The development of strategic reconnaissance technology was an important adjunct to the acquisition of nuclear weapons and long-range delivery vehicles. Historians and defense analysts generally agree that the advent of the latter heralded a revolution in military affairs, or, as Soviet theorists put it, a military-technical revolution.[41] In their view, the technology of nuclear weapons and associated delivery systems, combined with new operational concepts and organizations, changed significantly the character and conduct of warfare. The nuclear revolution created new ways of war and threatened to render existing ones obsolete.

Although the destructiveness of nuclear weapons was apparent at Hiroshima and Nagasaki, their revolutionary nature emerged more slowly. As late as 1951, nuclear warfare appeared to fit comfortably within the framework of pre–World War II strategic bombing theory. Theorists considered that atomic bombs were not so powerful that numbers and accuracy did not matter and armies and navies still had an important role to play. A future war would be one of attrition in which mobilization would be important.[42] However, between 1945 and 1960, as nuclear weapons became more powerful, plentiful, and deployable, how analysts thought about nuclear war changed.

The military's embrace of nuclear weapons was initially tentative. The actual capability of the United States to conduct nuclear operations remained quite limited in the years following World War II. In December 1945, the United States had three atomic bombs; in July 1946, it had nine; a year later it had thirteen; and a year after that it had fifty. All were Mark 3 "Fat Man" implosion devices, weighing five tons. None was assembled;

putting one together would have taken a crew of thirty-nine men two days. And it was not until 1948 that the Air Force had a fully qualified assembly team. The weapons were so large and heavy that they could only be mated to the bomber through the use of a special hoist. Through 1948, SAC had only some thirty specially modified B-29s capable of dropping the atomic bomb, all attached to the 509th Bomb Group at Roswell, New Mexico.[43] To make matters worse, in 1948 SAC had only some fifty crews trained to deliver nuclear weapons. When General Curtis LeMay assumed command of SAC he found that not a single crew was capable of delivering a weapon on target in anything approaching wartime conditions.[44]

One reason for the slow integration of nuclear weapons into the U.S. armed forces was the great secrecy with which they were treated. Atomic bombs were built and controlled by the Atomic Energy Commission (AEC), a civilian agency. The president was not formally briefed on the size of the nuclear stockpile until the spring of 1947, and the Joint Chiefs of Staff did not consider any war plan including nuclear weapons until late 1947. It was not until September 1948 that the National Security Council (NSC) approved NSC-30, "Policy on Atomic Warfare," which authorized the U.S. military to plan for the use of nuclear weapons in time of war. It also reserved for the President the authority to decide if and when they would be used.[45]

Between 1948 and 1952, a series of innovations allowed weapon designers to increase the nominal yield of the Mark 3 more than twenty-five times. These included advances in the design, composition, stability, and power of the high explosive charge used to create a critical nuclear mass as well as mechanics, structure, and composition of the fissile material itself. In 1951, the United States tested its first boosted atomic weapon, which used a small amount of fusion fuel within a hollow implosion core to produce a weapon with a yield approaching one megaton.[46] By 1952, the United States was mass-producing atomic weapons.

A coalition of civilian and military advocates, including Edward Teller and E. O. Lawrence and Atomic Energy Commissioner Lewis Strauss, urged the government to give top priority to the development of thermonuclear weapons. Such a project was, however, fraught with theoretical and engineering challenges. Nor was the scientific community united in its support of the project. J. Robert Oppenheimer, for example, wanted instead to perfect fission weapons. A brutal bureaucratic battle led to a decision in 1949 to initiate a crash program to develop a thermonuclear weapon, one Truman approved in January 1950.[47]

The United States tested its first true thermonuclear weapon in October 1952. The weapon—or, more accurately, the device—was a twenty-ton, cryogenically cooled behemoth with a yield of over ten megatons. However, a series of innovations allowed the United States to increase the yield and decrease the size of its weapons. By 1953 the SAC had an emergency capability to deliver the weapons. In addition, the development of "sealed pit" nuclear weapons technology permitted prolonged, safe, in-plane storage and transport of nuclear weapons and greatly enhanced readiness.[48] Changes in the design of the nuclear trigger and the adoption of extremely dense metal alloys permitted designers to fabricate nuclear weapons that used diminishing amounts of radioactive material. Smaller warheads translated into a larger stockpile and a growing variety of delivery means: not only ballistic and cruise missiles, but also depth charges, torpedoes, artillery shells, rockets, and mines. In addition, the Los Alamos and Lawrence Livermore National Laboratories discovered ways to reduce dramatically the size of nuclear weapons. Whereas the first thermonuclear weapon was twenty-two feet long, the nuclear warhead for the Davy Crockett mortar, fielded in 1961, was only two feet long and a foot in diameter.[49]

During the 1950s, nuclear strategy emerged as a separate field of academic study. Whereas serving or former military officers traditionally had an edge over their civilian counterparts in thinking through the problems of war, the advent of nuclear weapons leveled the intellectual playing field. As a result, in the 1950s it was civilian researchers, primarily at the Rand Corporation in Santa Monica, California, that developed deterrence theory, the concepts of first- and second-strike forces, and counterforce targeting. Indeed, nuclear strategists came from diverse backgrounds and included historians such as Bernard Brodie, economists such as Thomas Schelling, and mathematicians such as Albert Wohlstetter.

On October 29, 1953, President Eisenhower signed NSC 162/2, "Basic National Security Policy." The directive drew an explicit link between American national security and a healthy economy. In an effort to keep spending low, the U.S. armed forces would rely increasingly upon the threatened use of nuclear weapons. The centerpiece was the so-called doctrine of massive retaliation, the belief that U.S. security rested, in the words of Secretary of State John Foster Dulles, "primarily upon a great capacity to retaliate, instantly, by means and at places of our own choosing."[50] The document directed the Defense Department to arm each of the services with nuclear weapons. It called for the government to develop and deploy tactical nuclear weapons. It also recognized the need to protect the U.S. mobilization base though early warning and continental air defense.[51]

The Eisenhower administration's New Look defense plan was designed to implement the strategy of massive retaliation. The New Look reflected the belief that nuclear weapons had revolutionized warfare. It redefined the role of each service, aligning it with the atomic age. In so doing, the New Look significantly changed the size and composition of the services. During Eisenhower's two terms in office defense fell from 64 percent of federal spending to 47 percent. Manpower fell from 3.5 million to 2.47 million in 1960. Budget cuts forced the services to cancel many of their Korea-era expansion plans and eliminated six Army divisions, fifteen Air Force wings, and three hundred Navy ships by 1960.[52]

The New Look gave high priority to SAC as a mainstay of massive retaliation. SAC adopted new aircraft and weapons, expanded its base network, and improved its communication system. By contrast, the Eisenhower administration cut budgets for conventional ground forces substantially, betting that allies, backed by American air and sea power, would bear the brunt of fighting future regional wars.

The Eisenhower administration also saw a major shift in the size of the services. Eisenhower believed that air power held the key to deterrence. SAC's bombers and (eventually) ICBMs thus became the primary means of implementing massive retaliation. In FY53, at the end of the Truman administration, the Air Force's budget was slightly smaller than the Army's. Within two years, it had grown to nearly twice the size of the Army's budget. Indeed, the Air Force budget nearly equaled that of the Army and Navy combined.[53] In FY58, the Air Force took more than 48 percent of the budget, compared to 29 percent for the Navy and Marine Corps and 21 percent for the Army.[54]

In the early 1950s, atomic weapons were increasingly treated as conventional weapons. As Gordon Dean, chairman of the Atomic Energy Commission, wrote in 1952, "we can with complete justification treat the tactical atom—divested of the awesome cloak of destruction which surrounds it in its strategic role—in the same manner other weapons are treated."[55] Or, as General James Gavin put it, "Nuclear weapons will become conventional firepower.... To say that they will become conventional means that they will be in the hands of all military organizations including, for example, the smallest infantry units."[56]

Nuclear weapons posed both a threat to and an opportunity for the U.S. armed forces. On the one hand, they threatened the traditional identity of the services. On the other hand, the nuclear revolution opened up new areas of competition in ballistic and cruise missiles, space, and air defense. If a service could establish itself in these areas, then it could be assured of increased resources and prestige.

The Air Force: Exploiting the Nuclear Revolution

The Air Force was the primary beneficiary of the nuclear revolution. To many air power advocates, the advent of nuclear weapons seemed to validate the concept of strategic bombing that had animated aviators since the 1920s. Not surprisingly, the Air Force wholeheartedly embraced strategic nuclear bombing as its core mission. Bomber pilots dominated the Air Force as they had the Army Air Corps, and SAC became the most powerful organization in the service. The Air Force's embrace of strategic nuclear bombing yielded substantial dividends. During the 1950s, it garnered the lion's share of the defense budget. Nuclear-armed bombers, then nuclear-tipped missiles, became the coin of the realm.

The birth and growth of the Air Force were inextricably linked to the nuclear age. In March 1946, the U.S. Army Air Forces organized itself around three combat commands: the Strategic Air Command, Tactical Air Command (TAC), and Air Defense Command (ADC). A year and a half later, the 1947 National Security Act established the U.S. Air Force as a separate and equal member of the U.S. armed forces.

From the outset, SAC and its bomber pilots dominated the Air Force. SAC was a direct descendent of the 8th and 20th air forces, organizations that had waged strategic air warfare during World War II, and that experience shaped its postwar doctrine. The senior bomber generals believed that strategic bombing could not only win wars but also deter them.

In November 1946, SAC controlled two air forces and nine bomb wings. As one SAC press release put it, "Destruction is just around the corner for any future aggressor against the United States. Quick retaliation will be our answer in the form of an aerial knock-out delivered by the Strategic Air Command."[57] However, the command was more impressive on paper than in reality. Only six of SAC's wings had aircraft, the B-29 Superfortress. Only twenty-seven of these had been modified to carry the atomic bomb, all assigned to the 509th Bomber Group. The U.S. atomic stockpile contained thirteen weapons under the control of the Atomic Energy Commission. If given the order, it would have taken the bombers five days to pack up, move to an AEC depot, load the nuclear weapons, and deploy overseas.

In October 1948, the president named General Curtis LeMay commander-in-chief of SAC. LeMay was committed to attaining a high level of professional proficiency and readiness. Both would be required if SAC were ever called upon to launch a massive nuclear strike on the Soviet Union.

Strategic Bombers

In the early years of the Cold War, the bomber was the only platform capable of delivering nuclear weapons. The Consolidated Vultee Aircraft Corporation, or Convair, B-36 *Peacemaker* was SAC's first postwar bomber and first true intercontinental bomber. In fact, the B-36 was the result of an Army requirement, formulated in the spring of 1941, for a bomber that could take off from American territory, bomb Germany, and return home. It turned out to be a fortunate coincidence that the requirement for the long-distance payload of the B-36, developed before the advent of atomic weapons, equaled that of one atomic bomb (roughly ten thousand pounds) and its combat radius equaled the great-circle distance from Maine to Leningrad.[58]

The B-36 was a gargantuan aircraft; the *Enola Gay*, the plane that dropped the first atomic bomb, could nearly fit beneath one of its wings. Its massive bomb bay had the capacity of three railroad cars. Early models were powered by six turboprop engines; later models added four jet engines for takeoff, climbing to extreme altitudes, and dashing across hostile territory. With "six turning and four burning," the bomber could top 400 mph. On the other hand, the jet engines added weight and guzzled gas, reducing the bomber's combat radius to 3,110 miles.

The Peacemaker rapidly became enmeshed in bureaucratic battles between the Air Force and the Navy. At stake was not only budget share, but also roles and missions. Battles over the budget drove air advocate to extremes. The Air Force leadership and its allies in Congress hyped the bomber's range and payload, arguing that it was now possible to wage a war with a single weapon. As Secretary of the Air Force Stuart Symington told a New York audience in January 1948, "We feel, with deep conviction, that the destiny of the United States rests on the continued development of our Air Force. The question of whether we shall have adequate American air power may be, in short, the question of survival."[59]

Navy leaders, bitter over the cancellation of the supercarrier *United States* and concerned about the service's institutional future, publicly questioned the doctrine of strategic bombing and its embodiment, the B-36. The chief of naval operations, Admiral Arthur Radford, argued that the bomber had "become, in the minds of the American people, a symbol of a theory of war—the atomic blitz—which promises a cheap and easy victory if war should come.... Are we as a nation to have 'bomber generals' fighting to preserve the obsolete heavy bomber—the battleship of the air?"[60] The bomber's critics argued that the lumbering aircraft was vulnerable to Soviet interceptors and demanded tests of the bomber's survivability.

Nor were all the bomber's critics outside the Air Force. In 1947, the commander-in-chief of SAC, General George Kenney, argued that the B-36 was too slow to survive over enemy territory. Instead, he urged the Air Force to invest in bombers that could fly at the speed of sound, even if that meant shorter range and, hence, increased reliance on overseas bases. In the politically charged atmosphere of the day, many air power advocates within the Air Force interpreted his statement as disloyal, and he was fired.[61]

Kenney was, however, right about the bomber's vulnerability. In the late 1940s, the Soviet government embarked upon a massive strategic air defense program. Under Stalin the USSR spent far more on strategic air defense than on bombers and the atomic bomb combined. The result was a nationwide network of radar installations and command and control facilities. The Soviets fielded the MiG-15 jet fighter to intercept the B-29, eventually manufacturing over thirteen thousand of the aircraft. In 1957, the first Soviet surface-to-air missile (SAM) system, the S-25 (SA-1 Guild), became operational.[62]

In the end, the Air Force was forced to think of creative ways to try to increase the mammoth bomber's survivability. One idea was to equip the bomber with a pilotless drone to fight off enemy interceptors. Another was to outfit the bomber with a manned parasite—the XF-85 Goblin—that would ride in one of the B-36's bomb bays. Later, Republic adapted its F-84 Thunderjet to fit under the bomber's belly. The Air Force eventually considered converting the Peacemaker into a mothership that would linger offshore while the Thunderjet dashed in to take photographs or drop a bomb.[63] None of these schemes was ever put into effect, however.

The Air Force experimented with a number of novel follow-on aircraft designs. Perhaps the best known was the Northrop YB-49 flying wing, originally designed as a propeller-driven aircraft. In 1945, the Army issued a contract calling for the conversion of two airframes to jet propulsion. The aircraft entered testing in October 1947 and was found to consume a lot of fuel, have limited range, and be difficult to fly. The Air Force cancelled the program on March 15, 1950. In fact, the flying-wing design was an idea ahead of its time; it was reborn decades later as the design of the B-2 Spirit stealth bomber.

The B-52 Stratofortress was much more successful. The aircraft was developed in response to a requirement for a bomber capable of carrying a 10,000-pound bomb load five thousand miles. Dependable and easy to handle, the aircraft became the mainstay of SAC's bomber force throughout the Cold War and beyond. Indeed, B-52s saw heavy use, albeit in a much different role, during the Gulf War in 1991 and the campaign in Afghanistan in 2001.

The Defense Department's enthusiasm for all things nuclear was not limited to weapons. In 1946 the Joint Committee on Atomic Energy and the Atomic Energy Commission undertook the Aircraft Nuclear Propulsion Project, which was aimed at fielding a nuclear powered jet capable of extremely long continuous flight without refueling. In theory, a nuclear power plant could keep an aircraft in the air for days at a time. However, there was a great gulf between theory and practice. The highest hurdle had to do with the design of the reactor. A nuclear-powered aircraft required a small, light, yet powerful reactor as well as efficient shielding to prevent the crew from being irradiated. In the event, such a design proved to be out of reach, and the program was canceled in 1961.[64]

From Manned Aircraft to Cruise Missiles

Adapting to the advent of long-range ballistic and cruise missiles proved to be a greater challenge than nuclear weapons themselves. One the one hand, missiles appealed to the value the Air Force attached to technology in general, and the need to maintain technological superiority in particular. As General Hap Arnold testified before the Senate in October 1945:

> The first essential of air power necessary for peace and security is preeminence in research.... We must remember at all times that the degree of national security rapidly declines when reliance is placed on the quantity of existing equipment instead of its quality. We must count on scientific advances requiring us to replace about one-fifth of existing Air Forces equipment each year and we must be sure that these additions are the most advanced in the whole world. To this end the best scientific talents of the country must be mobilized continuously and without delay.[65]

As General White put it even more bluntly, "We in the Air Force... always want to see technology move faster because we realize that it is from the area of new developments that our lifeblood stems."[66]

At the same time, the development of long-range guided missiles called into question the central idea behind the Air Force—that of manned flight. Rather, they separated the operator from the vehicle he operated. When confronted with the need to adopt missiles, however, the Air Force understandably preferred the cruise missile, or "pilotless aircraft." It was only after an intercontinental cruise missile was demonstrated to be infeasible that the service began embracing the ballistic missile.

From the perspective of the late 1940s it was unclear whether the bomber or the missile would be the primary delivery means for nuclear weapons. On one hand, it was uncertain whether long-range missiles were feasible or practical. On the other hand, experts both inside and outside the government—including Theodore von Kármán, the director of the California Institute of Technology's Guggenheim Aeronautical Lab and the Army Air Force's Scientific Advisory Group, and mathematician John von Neumann—cast doubt on the future of the bomber.[67]

As of April 1946, the Army Air Force had twenty-eight missiles of all types in development. Some officers saw vigorous pursuit of missiles as vital to the Air Force's survival. As Major General Hugh J. Knerr, the secretary general of the Army Air Force Air Board, put it in February 1946, "The aerial missile, by whatever means it may be delivered, is the weapon of the Air Corps. Unless we recognize it as such and aggressively establish ourselves as most competent in this field, the responsibility therefore will become established by the Army or the Navy."[68]

Postwar budget constraints caused the Air Force to cut its missile program dramatically and drop all plans for ballistic missile development in July 1947. Of twenty-eight missile projects established a year earlier, only two survived: the Snark and Navaho cruise missiles. In choosing to emphasize cruise missile over ballistic missile development, Air Force decision makers made what they believed to be the safe bet. They felt that the evolutionary development of the cruise missile would yield an intercontinental weapon. Too few realized that a highly accurate five-thousand-mile ramjet-powered cruise missile presented a much more daunting challenge than a ballistic missile of comparable range.[69]

The U.S. Air Force's first "pilotless bomber," the Martin Matador, was designed as a tactical surface-to-surface missile. Development of the Matador began in August 1945 and the missile entered testing in January 1949. The missile was launched by a booster rocket from a mobile trailer and was controlled electronically from the ground during flight. It was capable of reaching a speed of six hundred miles an hour and had a range of six hundred nautical miles.

One of the Matador's weaknesses was its reliance upon command guidance, which required that the missile remain in contact with its ground station throughout its flight and thus limited its range. In an effort to extend the missile's range, the Goodyear Aircraft Corporation developed the Automatic Terrain Recognition and Navigation (ATRAN)—a radar map-matching system. The system correlated the return from a radar scattering antenna with a series of terrain maps carried on board the missile and which corrected the missile's flight path if it deviated from the map. First tested in the laboratory in March 1948, it was first flight-tested in October. Unlike command guidance, ATRAN could not easily be jammed and was not limited to the line of sight. It did, however, demand a library of radar maps for accurate navigation.[70]

In August 1952, Air Materiel Command combined the ATRAN navigation system with a modified version of the Matador airframe to create the TM-76A Mace. First deployed to Air Force units in Europe in 1959, it remained in service until the mid-1960s. The Air Force also developed the TM-76B Mace, guided by a jam-proof inertial guidance system and with a range twice that of the Mace A. Development of the Mace B began in 1959, it initially deployed to operational units in 1961, and remained operational in Europe and the Pacific until the early 1970s.

In March 1954, the Air Force deployed its first Matador unit in West Germany. Later units were sent to Korea and Taiwan. Eventually it deployed six squadrons with just under two hundred Matador and Mace missiles. However, the missiles suffered from low reliability and poor accuracy; the weapons were phased out as ballistic missiles entered the inventory. The Air Force deactivated the last Matador unit in April 1969.[71]

While short-range missiles such as the Matador and Mace were the first to see service, from the outset the Air Force had its eye on longer-range weapons. In January 1946, Northrop submitted a proposal to build a subsonic, turbojet-powered missile with a three-thousand-mile range, a program dubbed the MX-774A Snark.[72] It was designed to be launched from a mobile platform by two booster rocket engines that propelled the missile to flying speed before its turbojet engine started and the boosters were jettisoned. When it arrived over its target, its nose section, which contained the nuclear warhead, would separate from the fuselage and fall in a ballistic trajectory onto the target.

The challenges associated with developing an intercontinental cruise missile were many. Foremost among them was that of guiding the vehicle accurately over such long distances. Contemporary inertial navigation systems would drift too much during the missile's flight to provide the accuracy needed to deliver even a nuclear payload. Instead, Northrop proposed an innovative solution: a stellar guidance system that would match the missile's course to the position of the stars. The resulting system worked, but not reliably throughout the missile's flight. Moreover, it was large and weighed nearly one ton.[73]

The Snark was a controversial project. In July 1949, Air Force General Joseph McNarney touted it as America's most promising missile project. However, much of the Air Force was cool to the program. Perhaps predictably, the Army and Navy also criticized the program for its cost and risk.[74] The problems plaguing the Snark's test program fueled further condemnation of the program.

As if that were not enough, the Air Force soon leveled new requirements. Concerned about the survivability of an intercontinental cruise missile in the face of increasingly sophisticated air defenses, in June 1950 the program's managers added a requirement for a supersonic dash at the end of the missile's flight. They also increased its range, payload, and accuracy. Northrop's response was the Super Snark, a missile with a longer fuselage, sharper nose, and larger wing than the Snark. Like its predecessor, however, it suffered from technical problems, cost overruns, and schedule delays, all of which sapped support from the program.[75]

SAC began expressing its doubts in late 1951. Although some of this may have been the result of the dominance of bomber pilots within the Air Force, there were valid reasons to question the missile's survivability both on the ground and in the air. On the ground, it was to be deployed at vulnerable fixed sites. In the air, it was subsonic, unmaneuverable, and lacked defensive armament. The Strategic Missile Evaluation Committee, a panel of senior scientists convened to study America's long-range missile programs, found the Snark to be overly complex, vulnerable, and inaccurate. The missile had only a fifty-fifty chance of delivering its payload to within twenty miles, a level of inaccuracy that limited its effectiveness with even the largest nuclear warhead.[76]

Despite these problems, the Air Force moved forward with the program. The first Snark unit, the 702nd Missile Wing at Presque Isle, Maine, became operational in January 1959 and received its first missiles in May. President Kennedy scrapped the deployment shortly after coming to office two years later. The administration saw the missile as vulnerable on the ground, unreliable in flight, and unable to penetrate Soviet air defenses. More importantly, it had marginal effectiveness compared to ICBMs. The Air Force deactivated the 702nd Wing on June 25, 1961.[77]

Even more ambitious than the subsonic Snark was the supersonic Navaho. The missile was designed to redress one of the Snark's greatest weaknesses: its vulnerability in flight. The Navaho was designed to travel 3,500 miles at three times the speed of sound. Despite being a high priority for the Air Force, the missile was troubled from the start. Its ramjet engines proved particularly unreliable, earning it the nickname "Never-go." In early July 1957, after an expenditure of $700 million and less than ninety minutes of flight time, the Air Force cancelled the program and decided to rely upon the SM-62 Snark until the first-generation ICBM, the SM-65 Atlas, was deployed.[78]

The central place accorded manned aircraft in the Air Force shaped the way the service approached missiles. In the early years, the service tended to treat cruise missiles as manned aircraft. In May 1951, for example, the Air Force Council recommended assigning aircraft designations to guided missiles. The Matador, Snark, and Navaho were designated as bombers, while the Falcon air-to-air missile and Bomarc surface-to-air missile were designated as fighters. The Matador, for example, was designated the B-61 bomber. In addition, missile wings were included in the total number of Air Force wings.[79] Later, they were redesignated tactical (TM) and strategic missiles (SM), respectively.

Some—including members of the Air Force leadership—felt that the service was resistant to unmanned missiles. In their view, the Air Force was dominated by officers whom missiles threatened to put out of a job. Pilots had affection for the aircraft they flew and dismissed as misfits Air Force officers who were scientists.[80] In the words of Colonel Edward N. Hall, chief of the Western Development Division's propulsion development project, "The barrier to be overcome was not one of sound, or heat, but of mind, which is really the only type that man is ever confronted with anyway."[81] General Thomas D. White, who served as vice chief of staff and chief of staff of the Air Force between 1953 and 1961, was even more blunt. As he put it, "To say there is not a deeply ingrained prejudice in favor of aircraft among flyers would be a stupid statement for one to make. Of course there is."[82] At a 1957 commanders' conference, White likened the airman's attachment to the airplane to the cavalryman's attachment to the horse. Using blunt language, he castigated fellow officers for what he termed a "battleship" mentality. He told his subordinates that missiles were here to stay and that it was in the service's interest to get into the competition early. He also felt that once the value of the new weapons had been established, then they would gain acceptance.[83]

LeMay, for his part, saw missiles as adjuncts to, rather than replacements for, the manned bomber. As he put it, "I think it is reasonable to say that the first ICBM will augment the manned bomber force; and at some later date will supplant a portion of the manned bomber force. But I do not believe that in the foreseeable future the ICBM will replace all of the manned bomber force."[84]

From Cruise Missiles to Ballistic Missiles

A combination of technical effectiveness, threat perception, and bureaucratic politics drove the Air Force to develop ballistic missiles. Ballistic missiles—if they could be made to work—promised a number of advantages over bombers or cruise missiles. Foremost among these was their speed.

A ballistic missile would be able to reach its target in minutes, rather than hours for a bomber cruise missile. Just as important was the fact that they would be unaffected by Soviet air defenses. Whereas U.S. bomber bases were increasingly vulnerable to a Soviet first strike and airborne aircraft faced the growing challenge of penetrating Soviet air defenses to deliver their bombs, ballistic missiles offered an assured strike capability.

A number of respected scientists, however, remained skeptical. Vannevar Bush, who had served as the head of the Office of Scientific Research and Development, wrote in 1949 that "practical intercontinental missiles" were a "fantasy."[85] A 1947 Air Staff review panel predicted that for the next decade at least, subsonic bombers would remain the only way to deliver ordnance beyond a thousand miles.[86] Much of this skepticism was based upon uncertainty over the technical feasibility of an intercontinental ballistic missile, specifically the need to achieve accurate guidance, develop powerful rocket engines, and ensure the survival of a warhead as it reentered the atmosphere.[87]

Nor was there great enthusiasm for the ballistic missile within the armed forces. Although the services began experimenting with long-range ballistic missiles in 1945, none gave them a top priority. As far as the Air Force was concerned, surface-to-surface missiles fell behind air-to-surface and air defense missiles in priority.[88] And among surface-to-surface missiles, many Air Force leaders preferred the Snark and Navaho cruise missiles to ballistic missiles, even though neither demonstrated the ability to penetrate enemy air defenses.[89]

Counterbalancing such constraints were several powerful motivations to pursue ballistic missiles. First was the perception that the United States was falling behind the Soviet Union in the missile race. In late 1951 and early 1952, U.S. intelligence organizations received reports suggesting that the Soviets had developed a huge rocket engine capable of achieving 265,000 pounds of thrust, and that they had another design twice as powerful in development. This strongly suggested that the Soviets were developing an ICBM. If they were to deploy such a weapon, then they would be able to hold U.S. bomber bases at risk.[90]

Second, bureaucratic politics played a role. The Army had its own ballistic missile program, led by Wernher von Braun and his team of German scientists. Ballistic missile supporters within the Air Force were unwilling to cede such a promising area to the Army.

Third, several technological developments made the ICBM feasible. In December 1950, the Rand Corporation reported that significant advances in rocket engines and guidance had made long-range missiles possible.[91] The test of the first nuclear fusion device at Eniwetok in November 1952 demonstrated that it was possible to manufacture a nuclear warhead weighing as little as three thousand pounds, allowing missile designers to reduce drastically the payload that an ICBM would need to carry. The massive yield of the hydrogen bomb also reduced the requirement for an ICBM to deliver its payload with extreme accuracy. The latter development led the Air Research and Development Command to recommend the full-scale development of an ICBM.

Finally, intervention by Secretary of the Air Force Trevor Gardner and pressure by the Eisenhower administration helped convert the Air Force to support unmanned missiles. As Edmund Beard concluded in his path-breaking study of the development of the ICBM, had civilians not intervened, the Air Force would likely have continued to develop successive generations of manned bombers, despite doubts about their survivability.[92]

In 1950, the Air Council recommended the service establish a slow-paced project, dubbed Project Atlas, to study alternative designs for a ballistic missile with a range of five thousand miles and the ability to deliver a five-thousand-pound warhead to within five thousand feet of its target. The recommendation reflected the assumption that manned bombers would remain the backbone of strategic air power until 1965 and that increasingly more effective long-range cruise missiles would also enter the inventory. It would not be until the mid- to late-1960s that the Air Force would field ICBMs.[93]

Concern over Soviet missile developments and the inability of the strategic cruise missile soon caused the Air Force to change course. In May 1954 the secretary of the Air Force authorized "the maximum effort possible with no limitation as to funding" for the Atlas project. The following month it became the Air Force's most important program, one that would be "accelerated to the maximum extent that technological development will permit."[94] On July 1, 1954, General Thomas S. Power, the Air Research and Development Command (ARDC) commander, established an ARDC field office, the Western Development Division (WDD), under forty-three-year-old General Bernard Schreiver.

To field the missile as quickly as possible, WDD adopted concurrent and parallel development. Concurrent development entailed the simultaneous procurement of missile subsystems, manufacture and test facilities, command and control facilities, and training. For example, construction began on launch facilities for the Atlas before the engineers had finalized the missile's design.[95] Parallel development entailed hiring separate contractors for each subsystem. WDD also used competition to decrease timelines.[96] The Air Force granted the WDD's programs the highest priority and an unusual degree of freedom. By the end of 1955, WDD had established itself as a major weapon development center. Its projects included the Atlas and Titan ICBMs, Thor IRBM, and the WS-117L reconnaissance satellite.

Because no single company could muster the resources needed to act as the prime contractor for the Atlas, Schriever entrusted the Ramo-Woolridge Corporation to act as the systems integrator for the project. This was a significant innovation. Heretofore the prime contractor had been responsible for the development of a weapon. The ICBM was so complex that the Air Force needed a company to coordinate the work of the hundreds of companies working on the missile. Ramo-Woolridge became part of the Air Force family, with its engineers working side by side with Schriever's staff. Together they developed the discipline of systems engineering to coordinate the work of hundreds of contractors and the development of thousands of subsystems.[97] By 1957, the program involved seventeen principal contractors, two hundred subcontractors, and a workforce of seventy thousand.[98]

The early development of the Atlas was marked by a series of failures. Although the first missile was delivered in August 1956, its first successful launch did not occur until sixteen months later. The task of hurling a 3,000-pound nuclear warhead 5,500 nautical miles to within 1,500 feet of its target should not be underestimated. When initially fielded in 1959, Atlas Ds were deployed at aboveground gantry launchers at Vandenberg AFB, California. Later they were deployed in aboveground concrete "coffins." The missiles had to be fueled and raised before firing. Follow-on models of the Atlas were slightly less vulnerable. The inertially guided Atlas Es were deployed in earth-covered concrete coffins, while the Atlas Fs sat atop elevators inside underground concrete and steel silos.[99]

The biggest engineering challenge was the design of the missile's nose cone and reentry vehicle. At the time there was no proven way to reenter the atmosphere at high speeds and ensure the safe delivery of the payload to a point on the earth. Design of the reentry vehicle consumed as much as 11 percent of the Atlas program's budget.[100]

As a way of mitigating the risk inherent in the development of the Atlas, the Air Force used the same components to develop the XSM-68 or WS107A-2 Titan. Like the Atlas, it was a liquid-fueled ICBM guided by an inertial navigation system. The Titan I was the first U.S. Air Force ICBM to be placed in a hardened underground silo for protection against enemy attack. First tested in February 1959, it was eventually deployed at five bases in the western United States. By 1965, however, it was being phased out in favor of the Titan II.

The period from 1957 to 1963 marked the transition from bombers and cruise missiles to ballistic missiles. As Robert Perry has noted, before 1957, ballistic missiles were handicapped in their competition with bombers and cruise missiles; by 1963, however, they had become the chief instruments of strategic warfare for the United States. Such a shift was manifest in the Air Force's acquisition decisions. Between 1951 and 1962, nearly three thousand jet-powered strategic bombers entered the inventory. By the end of 1962, however, it was clear that relatively few would be replaced after they became obsolescent. The ballistic missile had become the dominant weapon.[101]

In January 1955 the Air Force Scientific Advisory Committee recommended developing a tactical ballistic missile. As with the Air Force's ICBM program, the intermediate range ballistic missile (IRBM) was driven by a mixture of threat and bureaucratic interest. The Killian Committee recommended that the United States develop a tactical ballistic missile before the Soviets. Moreover, Army was developing the Jupiter missile. The Air Force's response to both challenges was the Thor.[102]

Development of the Thor was rapid, with the first test missile delivered to Cape Canaveral less than a year after the development contract was signed. The missile entered service in September 1958. Because it could not strike the Soviet Union from U.S. territory, the United States entered into an agreement with Great Britain to deploy the missiles in the British Isles. The Air Force furnished Thors to Britain and trained the crews, and Britain agreed to build bases and man the missiles. The United States deployed four squadrons of fifteen missiles each in Britain. It also deployed two NATO squadrons of Jupiter missiles in Italy and one in Turkey.[103]

The move from bombers to missiles changed not only the U.S. defense posture, but also the structure of the U.S. defense industry. General Motors Corporation, which had been the largest government contractor during World War II, had fallen to twenty-first place by 1960. Curtiss Wright dropped from second to thirtieth place, Ford from third to thirty-sixth place, and Bethlehem Steel from seventh to forty-second place. By 1960, these industrial giants had been replaced as the top five defense contractors by Boeing, General Dynamics, Lockheed, General Electric, and North American Aviation.[104]

Continental Air Defense

The growth of Soviet nuclear attack capability throughout the 1950s raised the possibility that the Soviet leadership could order a surprise attack on the U.S. bomber force. SAC attempted to protect its bombers through dispersal and defense. In 1951 the Air Force created the Air Defense Command, an integrated system of interceptor aircraft, anti-aircraft missiles, and early warning radar installations. The following year, President Truman ordered the Air Force to construct the Distant Early Warning (DEW) line across the top of North America to provide warning of a Soviet attack.

The need to protect SAC's bombers and America's cities from nuclear attack pushed the Air Force to think seriously about continental defense. To remain effective in the face of a large, sophisticated attack, such a defense would need to be automated. The Air Force developed the Semi-Automatic Ground Environment, or SAGE, system to defend against Soviet long-range bombers. SAGE, developed by MIT's new Lincoln Laboratory and its spin-off, the Mitre Corporation, revolutionized air defense. The project saw the development of the first real-time control computers, the AN/FSQ-7 and AN/FSQ-8. Each contained 25,000 vacuum tubes and 147,456 ferrite cores. Installed in pairs, the computers weighed 275 tons each, occupied 40,000 square feet of floor space, and used 3 million watts of power.[105]

Initially designed to protect the northeastern United States from Soviet bombers, SAGE would eventually cover the entire continental United States. The United States was divided into eight SAGE sectors, each with a combat center equipped with an AN/FSQ-8 computer, tied directly to the North American Aerospace Defense Command (NORAD) Combat Operations Center. Each combat center would receive, process, and evaluate information from radars in thirty-two subsectors, each with a direction center with an online AN/FSQ-7 and an identical standby. Combat centers would generate an overall view of threats and responses, while direction centers would process surveillance information and assign aircraft and missiles to threats. Direction center computers were designed to track two hundred enemy aircraft as well as two hundred defensive aircraft and missiles simultaneously.[106]

A masterpiece of technological sophistication, SAGE was a military failure: By the time it reached initial operational capability in 1959, Soviet ICBMs had replaced bombers as the greatest threat to the United States. On the other hand, the project yielded valuable spin-offs in computers, communications, and management techniques.[107] Moreover, it remained in operation as part of NORAD's attempt to defend the continental United States against Soviet bombers.

The Nuclear Challenge to the Navy

Whereas the nuclear revolution benefited the Air Force, the advent of nuclear weapons and long-range missiles led many to call into question the continued relevance of the Navy, which met the challenge in several ways. The Navy originally chose incremental adaptation by fielding nuclear bombers on its aircraft carriers. With the development of the fleet ballistic missile (FBM) and the nuclear-powered submarines to carry and deliver them, however, the Navy staked its future on the submarine-launched ballistic missile (SLBM). In so doing, the Navy found a way of adapting to the nuclear age in a manner consistent with its organizational culture.

The emergence of the Air Force—and SAC in particular—as the premier arm of the U.S. military elicited a range of responses from within the Navy. Although some Navy officers wanted to beat the Air Force at its own game by transforming the Navy into a strategic Air Force, most believed that it was more important for the Navy to demonstrate its continued relevance by showing that its ships were not excessively vulnerable to atomic attack and that Navy carrier aircraft were at least as useful as Air Force bombers for nuclear strike.[108]

Several younger officers, including Commander Frederick L. "Dick" Ashworth and Commander John T. "Chick" Hayward, were more convinced of the need for the Navy to demonstrate its worth in nuclear terms. In their view, atomic weapons would enhance the Navy's ability to perform existing missions such as carrier-based strike.[109]

The key to demonstrating the Navy's utility was finding a suitable combination of ship, aircraft, and bomb. Whereas the Air Force was able to adapt existing strategic bombers to nuclear delivery, the Navy had to look to new aircraft for the nuclear attack mission. North American Aviation was already building the AJ-1 Savage attack aircraft. In 1946, the Navy let a contract to modify the aircraft to carry an atomic bomb.[110] In the interim, the Navy modified a dozen P2V Neptune ASW patrol bombers to carry nuclear weapons. The aircraft were stripped of all expendable equipment to lighten their weight and installed with tailhooks to allow them to land on aircraft carriers. Originally designed to take off from land bases, the aircraft were so heavy they had to be loaded aboard the carriers by crane while at dockside and needed jet assistance to take off from the carrier's deck.[111] Similarly, the flight deck of the carrier had to be strengthened to handle the heavy bombers.[112]

In the first sortie of a nuclear-capable carrier aircraft, on April 27, 1948, Hayward flew a P2V off the USS *Coral Sea*. By December, the Navy commissioned a developmental squadron under the name Composite Squadron Five (VC-5) with Hayward as Commanding Officer. In January 1949, part of the squadron was split off to form VC-6 under Ashworth. In September, Hayward demonstrated the P2V's capabilities aboard the *Midway* before an audience that included Secretary of Defense Louis Johnson,

Secretary of the Air Force Stuart Symington, Secretary of the Army Gordon Gray, Secretary of the Navy Francis Matthews, and Chairman of the Joint Chiefs of Staff Omar Bradley. In February 1950 the Navy deployed its first nuclear-capable squadron overseas when VC-5, equipped with the new AJ bombers, deployed.[113] Despite such "firsts," the Navy possessed a limited nuclear capability: It was not until later in 1950 that an AJ-1 even attempted to take off from or land on an aircraft carrier. Moreover, the aircraft was a major failure. It was not until the deployment of the A3 Skywarrior in the second half of the 1950s that the Navy possessed an effective nuclear bomber.

The Navy also explored other options for using aircraft to deliver nuclear weapons. In the early 1950s, the Navy and defense industry explored the possibility of using nuclear-armed seaplanes to form a Seaplane Striking Force. Operating from dispersed remote sites supported by ships and submarines, squadrons of seaplanes, including the Convair XF2Y-1 Sea Dart and the Martin P6M SeaMaster, would conduct antisubmarine warfare, mining, and launch strikes against the Soviet homeland. However, the concept lacked institutional support, as the seaplane community represented a minority within the aviation community, which was itself only one of three main communities within the Navy. As a result, it enjoyed a low funding priority relative to carrier aviation and submarines.[114]

The Navy and Cruise Missiles

The Navy, like the Air Force, paid considerable attention to missiles in the years following World War II. The Navy's need for a nuclear delivery system, combined with interservice competition with the Air Force, drove the quest. In the late 1940s it conducted experiments with ship- and submarine-launched cruise and ballistic missiles. For example, in 1947 the Navy experimented with launching an adaptation of the German V-1 cruise missile, the JB-2 Loon, from a converted Gato-class diesel submarine.

Such early efforts led to the development of the Chance Vought Regulus submarine-launched cruise missile.[115] The Regulus looked and performed a lot like the Air Force's Matador. First launched in July 1953 from the submarine USS *Tunny*, it relied upon two other submarines to guide it to the target. Later, using the so-called Trounce system, a single submarine could guide the missile to its target.[116]

Declared operational in 1955, the Regulus could also be launched from surface ships. Indeed, cruisermen were enthusiastic about the missile because it promised to extend their offensive range. A later version was capable of carrying a 3.8-megaton thermonuclear warhead 575 miles at Mach .87. A follow-on, supersonic version of the missile, the Regulus II, was cancelled owing to its cost and its unattractiveness compared with other means of delivering nuclear weapons.[117]

The Navy and Ballistic Missiles

The Navy was the last of the services to embrace the ballistic missile. As Vincent Davis has argued, the Navy lagged because "there were many naval officers who generally opposed a new emphasis on missiles, some because they questioned the Navy's need for missiles, some because they questioned whether basic scientific and technological research had made enough progress to warrant a new emphasis, and some because they were apprehensive that a new emphasis on missile development would mean a decreased emphasis on other high priority Navy programs at a time when naval appropriations were still relatively restricted."[118]

The barriers to producing a SLBM were high. There was no proven nuclear weapon design of sufficient yield that was small enough to be carried aboard a submarine, no accurate guidance system, and no solid propellant energetic enough to propel the missile. Indeed, there was no guarantee that any amount of money would be enough to bring such a missile into existence.

Because of such concerns, in the summer of 1955, the chief of naval operations (CNO) decided that moving ahead with the FBM was premature and cancelled the program. At the same time, Deputy Secretary of Defense Ruben Robertson announced a tentative decision to give the Air Force a monopoly over IRBMs, a move that would have shut the Navy completely out of the ballistic missile arena.[119]

The FBM program gained a new lease on life with the appointment of Admiral Arleigh Burke as Chief of Naval Operations in August 1955. Within twenty-four hours of taking office, Burke called for a briefing on the FBM program; within a week he had decided to revive the program and took a personal interest in it.[120] In 1955 he formed a Special Projects Office under Rear Admiral William F. Raborn and gave him license to recruit the fifty best people he could find.[121] Burke received additional backing when the final Killian Committee report was released in September 1955. The report recommended that the Navy develop a 1,500 mile sea-based missile to provide the United States with a secure retaliatory capability.[122]

The resulting FBM program was the first development effort of its kind. It not only produced the first SLBM, the Polaris A-1, but also the first nuclear-powered ballistic missile submarine (SSBN), the USS *George Washington* (SSBN 598). And it did so in a remarkably short period of time: On July 20, 1960, *George Washington* successfully fired two Polaris A-1 SLBMs while submerged off the coast of Cape Canaveral, Florida.

As Harvey Sapolsky has written, competition with the Air Force and its ICBM program helped drive the Navy to accept Polaris. This interservice rivalry cleared away bureaucratic obstacles and helped Polaris program managers assemble talent and garner resources, which in turn allowed them to field the missile ahead of time and within budget.[123]

High accuracy was not a key design parameter for the Polaris. There were clearly limits to how accurately a missile launched from a moving platform located in the middle of the ocean could find a target thousands of miles away. Moreover, its designers assumed that the missile would be used against large targets such as urban-industrial complexes. Even so, Polaris posed a significant guidance challenge. In order for the submarine's missiles to reach their targets, they would first need to be oriented to fly in the right direction. The Navy took extraordinary measures to determine the location of SSBNs. This included equipping the boats with Ships Inertial Navigation System, or SINS; conducting detailed surveys of the ocean floor in SSBN patrol areas to allow the boats to use their sonar to reset SINS; and information from the ground-based Loran-C or space-based Transit navigation networks. Moreover, the Navy developed a new, more accurate model of the earth's gravitational field to help subs determine their location.[124] The combined output of the submarine's navigation and fire control systems and the missile's guidance system would bring the missile to the vicinity of its target.

The Polaris A-1 missile carried a single 600-kiloton nuclear weapon and had a range of approximately 1,000 nautical miles.[125] The 1,500-nautical-mile Polaris A-2 was quite similar to the A-1; its increased range was the result of reduced weight and increased thrust in the missile's second stage. However, it was armed with a more powerful 800-kiloton warhead. Its first successful submerged launch came from the USS *Ethan Allen* (SSBN 608) on October 23, 1961. Eight months later, the Polaris A-2 began its initial operational patrol aboard the same submarine.

The Polaris A-3, deployed in 1964, was the first SLBM to have a range of 2,500 nautical miles as well as the first SLBM to have multiple reentry vehicles: it carried three 200-kiloton warheads. The missile was basically a new design, rather than an evolution of the A-1 and A-2. The range of the Polaris A-3 gave the United States the ability, for the first time, to cover the entire Eurasian landmass from submarines offshore.

Nuclear-Powered Submarines

The nuclear revolution yielded not only weapons but a new means of propulsion as well. Nuclear submarine propulsion solved several problems for the Navy. First, it allowed SSBNs to stay on patrol, submerged and unprotected, for long periods of time. As a result, it turned the Navy's nuclear arm into a hedge against a Soviet strike on U.S. ICBMs and bombers. Second, it gave nuclear-powered attack submarines (SSNs) the endurance they would need to operate far from U.S. shores.

The Navy's move to adopt nuclear submarines was rapid. In 1949, the Navy decided to go ahead with a project to develop a nuclear-powered submarine. On January 17, 1955, the first such boat, the USS *Nautilus*, departed the Electric Boat Shipyard in Groton, Connecticut, on atomic power. Nine months later, Burke declared that that all future submarines would be nuclear-powered.[126]

The Navy readily accepted nuclear submarines because they fit comfortably within the identity of the submarine community. Indeed, Owen Coté has termed the nuclear submarine a "true submarine," "one that needed no umbilical cord to the surface and could remain completely submerged."[127] They never had to snorkel and were so fast that active sonars could not keep their beams focused on them. Moreover, their speed and three-dimensional maneuverability allowed them to outrun existing torpedoes, which were designed to attack slower diesel submarines. The nuclear submarine possessed all the qualities that submariners had long wanted but did not possess.

The speed and endurance that nuclear propulsion permitted spawned other innovations. The USS *Albacore* (AGSS-569), commissioned in 1953, tested the use of the teardrop-shaped hull and a single screw to maximize the submarine's submerged speed. It was also the first U.S. submarine to use high-yield steel in its hull. These design features, among others, led to the deployment in 1960 of the Thresher (SSN-593) class, the world's first class of quiet nuclear submarines. Indeed, all subsequent classes of U.S. attack submarines derive from the Threshers. The boats were designed explicitly to hunt down Soviet nuclear submarines. The key to their effectiveness was quieting. The class's design used rafting, whereby the boat's engineering plant was placed on a flexible mount, or raft, within the submarine, to reduce the amount of noise it produced.[128] Such practices gave them a significant noise advantage over their noisy Soviet prey.

The transition to a submarine force dominated by nuclear boats was rapid. On January 1, 1961, there were 115 diesels, but only thirteen nuclear-powered submarines in the U.S. inventory. By the end of 1975, the ratio was almost completely reversed, with 106 nuclear-powered submarines and only twelve diesel submarines in the fleet.

The Army: Weathering the Storm

The advent of nuclear weapons threatened the Army more than any other service. In the years that followed World War II, the service faced the challenge of adapting its organization and doctrine to rapid technological change. It responded by adopting nuclear weapons for land warfare and even competing with the Air Force in the development of long-range missiles, space, and strategic air defense. It also undertook a radical—though ultimately unsuccessful—restructuring of its forces in a bid to retain its relevance on the nuclear battlefield.

The Army was the biggest loser in the organizational and fiscal battles brought on by the development of nuclear weapons. It ended the Korean War with twenty combat divisions; by 1961, it had been reduced to fourteen divisions, including three training formations. Throughout the Eisenhower administration, the Army enjoyed the smallest share of the defense budget of any service.[129] One officer believed that the Army had been reduced to the status of "an auxiliary service."[130] As General John H. Cushman candidly admitted in 1954, "I do not know what the Army's mission is or how it plans to fulfill its mission. And this, I find, is true of my fellow soldiers. At a time when new weapons and new machines herald a revolution in warfare, we soldiers do not know where the Army is going and how it is going to get there."[131]

The advent of nuclear weapons and strategic air power appeared to call into question the utility of traditional ground forces. At the very least, it seemed to demand a fundamental reconsideration of Army weapons, doctrine, and organization. As John K. Mahon wrote in 1954, "It may be that atomic power coupled with air power has changed [the role of armies]. So lethal a combination may at last have altered the basic role of land armies. No man can be sure. It is certain, however, that the experience of the last war cannot be relied on to any great extent in preparing for the next (should the nations be foolish enough to permit one to start)."[132]

In such an environment, a group of generals led by Matthew Ridgway, who had performed superbly as commander of the Eighth Army in Korea, spearheaded an effort to transform the Army. Although the Army leadership rejected the premises of the New Look, with its assumption that nuclear weapons and long-range air power would be the primary instruments of deterring (and if necessary fighting) future wars, they nonetheless embraced technology as the principal determinant of how wars would be fought. At the heart of this approach was the belief that although strategic nuclear weapons were insufficient to guarantee American security, tactical nuclear weapons would be sufficient to decide future wars.[133] Strategic nuclear weapons, they believed, were too destructive to be useful; their utility was confined to deterrence. Tactical nuclear weapons, by contrast, could be used to good effect on the battlefield without fear of escalation.

Some saw the development of nuclear weapons as changing the character of conflict. As two Army officers wrote in 1958, "The advent of atomic weapons on the battlefield has produced a revolution unequalled in military history."[134] More common, however, was the view that tactical nuclear weapons offered greatly improved firepower.

Tactical nuclear weapons comported with the Army's historical reliance on firepower. In many ways, the Army was predisposed to nuclear weapons. There was a good fit between nuclear weapons and the Army's tradition of substituting technology for manpower. The Army viewed tactical nuclear weapons, not so much as small strategic bombs, but as very powerful artillery.[135] To many Army officers, nuclear weapons were the ultimate expression of battlefield firepower. As General Willard G. Wyman, the commander of Continental Army Command, put it, thanks to nuclear weapons "tactical firepower alone can now accomplish the purpose of maneuver."[136]

At the same time, the idea that technology was a critical element of war ran counter to the Army's belief that the soldier stood at the center of battle. Not all Army officers believed that nuclear weapons provided an absolute guarantee of victory. Major Marvin Worley voiced this view when he wrote in 1959, "Many senior Army officers do not subscribe to the theory that there is an ultimate weapon, and certainly don't subscribe to the theory that an intercontinental ballistic missile is such a weapon."[137]

The Army had bureaucratic motives for pursuing nuclear weapons. The Eisenhower administration and Congress both showed greater enthusiasm toward nuclear arms than traditional weapons. To the extent that the Army needed to justify its budget, it was in nuclear terms. As Major General John B. Medaris, head of the Army Missile Office, put it, "If you put all your energy and effort into justifying these conventional weapons and ammunition... I think you are going to get very little money of any kind. It is far easier to justify a budget with modern items that are popular.... Why don't you accentuate the positive and go with that which is popular, since you cannot get the other stuff anyway?"[138]

The Army initially attempted to fit nuclear weapons into its traditional organizational culture and force structure. The service's first nuclear program focused upon fielding an atomic round for a 280mm cannon, the smallest cannon that could fire an atomic projectile. The cannon was immense: 85 feet long, it weighed 50 tons in firing position and 86 tons on its transporters, and had a maximum road speed of 35 miles per hour.[139] In May 1953, the Army fired its first nuclear projectile at the Nevada Test Site. Within months it had deployed six of the massive cannons to Europe.[140]

An evolutionary development of traditional artillery, the 280mm atomic cannon possessed none of the qualities the Army needed. It was road-bound and cumbersome, and its seventeen-mile range gave it little ability to reach deep targets. To strike beyond the front lines it would have needed to be deployed far forward, where it would have been vulnerable to attack and capture.

Ballistic Missiles and the Army

Rockets and guided missiles represented the most exciting field of military technology in the 1950s. Not surprisingly, missile development became a source of heated inter-service rivalry. The Army investigated missiles for space exploration, long-range attack, and air defense. Rather than ceding the field to the Air Force, it competed vigorously in each area. Indeed, it managed to elbow its way into each area.[141]

The Army's missile program was designed to provide nuclear fire support to Army formations. The program was remarkable for the variety of systems that it developed, including the Honest John and Little John battlefield rockets, Corporal and Sergeant battlefield missiles, and Redstone and Jupiter missiles.

The seventy-five-mile, liquid-propelled Corporal was the first guided ballistic missile in the U.S. arsenal.[142] Developed by the California Institute of Technology's Jet Propulsion Laboratory beginning in 1949, it had a length of forty-five feet and a thirty-inch diameter and could carry either a nuclear or conventional warhead. The nuclear-armed variant provided enormous firepower: Four Corporal battalions exceeded the firepower of all American artillery units in World War II.[143] By the end of 1957, some nine hundred Corporals had been produced. The U.S. Army deployed the missiles to Germany, Italy, and Great Britain and sold some to the British Army.

The Army developed the Sergeant as the replacement for the Corporal.[144] With a range of eighty-four miles, it was designed to attack military bases, airfields and concentrated military units with a nuclear warhead. It was more powerful, accurate, and reliable than the Corporal, featured a simpler, solid-fuel design and had an inertial guidance system. It was first deployed in September 1962 and remained in the inventory until it was replaced by the Lance beginning in 1973.

The Army's leadership wanted nuclear capabilities integrated into all levels of operations. First tested in August 1951, the Honest John was organic to airborne, infantry, and armored divisions; it was also deployed with Army missile commands.[145] The Honest John was a 762mm solid-fuel, unguided rocket with a twenty-two-mile range. The basic M31 design was deployed in 1954. Beginning in 1961, it was replaced by an improved version that was shorter, lighter, and had longer range.

In 1956, the Army began developing the Little John, a smaller version of the Honest John designed for airborne units. The 318mm, solid-fuel unguided rocket with a ten-mile range and its launcher were light enough to be transported by helicopter.[146]

The Redstone missile grew out of a U.S. Army Ordnance Department research and development contract with the General Electric Company for the development of guided missiles.[147] The project, known initially as Hermes, soon became known as Project Redstone. The 69-foot, 40,000-pound Redstone missile that resulted was equipped with an inertial guidance system and propelled by a mixture of liquid oxygen, alcohol, and hydrogen peroxide, giving it an operational range of 240 miles. First tested in May 1953, by 1956 it had been deployed in its first operational unit, the 40th Field Artillery Missile Group at Redstone Arsenal, Alabama.[148] The missile, designed to support a field army, was deployed with heavy missile commands.

The Redstone proved to be quite versatile. It was used not only as a missile system, but also as the first stage of the launch vehicle for *Explorer I*, the first U.S. satellite. And the Mercury Redstone carried first a chimpanzee, Ham, and then Navy commander and astronaut Alan B. Shepard on their suborbital flights.

Driven by competition with the Air Force for the intermediate-range nuclear mission, in 1955 the Army launched a crash program to develop the 1,500-mile Jupiter IRBM. It had no clear antecedent; it was neither a refinement of a fixed-wing aircraft, nor an evolution of traditional cannons or rockets. As a result, both the Army and Air Force could lay claim to the development of IRBMs and the resources that went with the mission. As a result, the Army sought to accelerate the Jupiter program and delay a decision on roles and missions as long as possible.[149]

The development of the Jupiter triggered a dispute with the Air Force, which was developing its own IRBM, the Thor. The Army justified the Jupiter on the grounds that it could not count on the Air Force to hit deep targets and that it provided night and all-weather coverage of targets. In November 1956, Secretary of Defense Wilson gave the Air Force jurisdiction over all IRBMs. He still allowed the Army to deploy the Jupiter, but only if the missile were controlled by Air Force personnel. He also restricted the range of future Army surface-to-surface missiles to two hundred miles.[150]

The Army's development of nuclear weapons came at the expense of more traditional weapon systems. In FY 1957, for example, the Army spent 43 percent of its research and development budget on nuclear weapons and missiles but gave only 4.5 percent to new vehicles, 4.3 percent to artillery, and 4 percent to aircraft. It devoted little funding to armor. The T113 armored personnel carrier (APC) spent much of the decade in research and development. It remained underfunded at the end of the decade.[151]

Strategic Air Defense

The Army and Air Force also clashed over the mission of continental air defense. The 1948 Key West agreement, which delineated the roles and missions of the U.S. armed forces, gave the Air Force the mission of defending the United States against bombers, but also specified that one of the Army's "primary functions" was to organize, train, and equip air defense units. The Army exploited this ambiguity to launch a massive continental air defense program, one that envisioned the formation of 150 air defense battalions to protect major cities.[152]

The Army's continental air defenses originally relied upon the Skysweeper radar-directed automatic cannon. The growth of the Soviet bomber threat soon called for a more sophisticated approach. A competition between the Army's Nike Ajax and Air Force's Bomarc surface-to-air missiles led to the deployment of the Ajax in 1954. The 21-foot, liquid-fueled supersonic missile relied upon command guidance to intercept its target.[153] The Army rushed the missile into production in an effort to meet the Soviet bomber threat, deploying ten thousand of the missiles at a cost of nearly $2 billion around key urban, military, and industrial locations. For example, sixteen Nike batteries protected Los Angeles, while twenty guarded New York. Nikes even ringed cities far inland, such as St. Louis and Omaha.

Despite high public support for national defense, such missile installations were not popular. A series of accidents further reduced the popularity of the system. In April 1955, a Nike was accidentally launched from a battery at Fort Meade, Maryland. The missile broke up and fragments fell onto the Baltimore-Washington Expressway, but fortunately nobody was hurt. An accident three years later produced more tragic results. In May 1958, eight Nikes exploded or burned on the ground at a battery near Middletown, New Jersey, killing ten and injuring three.[154]

The Nike Ajax's main limitation was its ability to engage only one target at a time, as well as the inability to coordinate multiple batteries. These shortcomings raised the prospect that a Soviet bomber attack would swamp U.S. air defenses. To alleviate this problem, the Army Air Defense Command established centers where incoming targets were plotted manually and engagement orders passed to batteries. Such a system proved inadequate, however. In the late 1950s, the Army introduced the Interim Battery Data Link to allow batteries to share data in real time.

Although the Nike Ajax had been an important step, its limited twenty-five-mile range and small conventional warhead sharply curtailed its utility. The Army explored outfitting the Ajax with a nuclear warhead, but that proved impractical. As a result, in July 1953 the Army authorized the development of a second-generation SAM, the SAM-N-25 Nike Hercules with a longer range and a forty-kiloton nuclear warhead. The missile was able to intercept targets at three times the range of the Nike Ajax. Capable of traveling at Mach 3.5 and intercepting high-altitude targets, it could defend an area of about twenty thousand square miles.[155] By June 1958, it was deployed at converted Nike Ajax sites near New York, Philadelphia, and Chicago. The Army eventually deployed 145 Nike Hercules batteries, including 35 built for the new missile and 110 converted Nike Ajax sites.

The deployment of the Nike Hercules exacerbated the dispute between the Army and Air Force. The Air Force charged that the Nike Hercules duplicated the capabilities of its soon-to-be-deployed Bomarc missile. In the end, both missiles were deployed, although the Nike Hercules was fielded in far greater numbers.

The task of protecting the United States from Soviet bombers was a complex one. The need to develop an effective integrated air defense system drove the development of some of the first computers. The Army, for example, developed the Missile Master to control and coordinate Nike air defense batteries. The Missile Master was the first truly integrated command and control system utilizing automatic data communications, processing and display. Installed at Army Air Defense Command posts, the system collected information on the location of aircraft, identified them, presented information, and distributed it to missile firing batteries. Missile Master operators could monitor and direct all air-defense batteries in an area to engage up to twenty-four targets. The focal point of the system was the Anti-Aircraft Operations Center, which processed information from

Army radar sets and the Air Force's SAGE interceptor control system and passed it along to air defense batteries.[156]

The Pentomic Army

The advent of nuclear weapons forced Army leaders to revise radically their view of warfare. Nuclear weapons would force armies to achieve greater dispersion, security, and deception in order to survive on the battlefield. Rather than mass, soldiers would need to disperse in breadth and depth in order to lessen an adversary's incentive to use tactical nuclear weapons. The challenge was developing the capability to mass forces at the right time and place, deliver a decisive blow, and then disperse.[157] As James Gavin put it, the challenge was to learn "how to control the amorphous mass of men who must be dispersed over an entire zone, an entire tract of land, dispersed thinly enough not to invite bomb blast, yet strongly enough to tackle the enemy."[158]

The advent of nuclear weapons thus created a need for mobile, hard-hitting combat organizations tailored to fight and survive on the atomic battlefield. In April 1954, General Ridgway directed the Army to develop organizations that were more mobile and flexible, that exploited new technology, and that could disperse to avoid nuclear effects. In November 1954, he commissioned a second study of Army organization, the Pentana study. In June 1955, General Maxwell D. Taylor succeeded Ridgway as chief of staff. He took an intense personal interest in the Pentana study.

Taylor outlined the new organizational structure, the pentomic division, in October 1956. In search of units that were capable of fighting independently yet were expendable, the Army moved from triangular organizations with three subordinate commands to pentomic organizations with five subordinate commands. In pentomic units the Army replaced its battalions with battle groups, each of which was to be capable of independent operation. The pentomic infantry division, for example, was composed of five infantry battle groups, an armored battalion, and a cavalry squadron. A transportation battalion controlled armored personnel carriers. The Army also formed pentomic airborne divisions, equipped almost entirely with equipment that could be transported by air.[159] The design of the armored division, by contrast, changed relatively little.

The nuclear age led the Army to reevaluate other areas of technology as well. Some officers felt that the best way to survive a nuclear attack was to dig in or disperse forces. As two of them wrote in 1958, "We must produce a device which will permit an individual to dig a deep foxhole in a matter of minutes, so that a unit could disappear underground as quickly as those sand crabs which live on the edge of a beach."[160]

Others sought to increase dramatically the mobility of conventional ground forces. They were particularly interested in technologies to increase the speed, range, and precision of ground forces. The Army explored "universal vehicles" capable of fast cross-country speed and road mobility that could also fly by means of modified rotors.[161] It pursued a number of tilt-rotor aircraft, including the Bell XV3 and Vertol VZ2. It also investigated "individual lift devices" designed to move a single soldier safely over the nuclear battlefield. One such design, the De Lackner Aerocycle, was a platform equipped with a 43-horsepower engine and two counter-rotating blades designed to fly at 65 miles per hour with a 150-mile range. Another, the Hiller Flying Platform, was a cylinder housing three engines that powered two counter-rotating propellers. The Army also let contracts to Chrysler, Piasecki Aircraft Corporation, and Aerophysics Development Corporation to develop a jeep that could hover and fly.[162] Even more exotic ideas included disposable uniforms of "non-woven film," maintenance-free trucks that would be driven a thousand miles and then discarded, and the use of cargo rockets for battlefield supply.[163]

Still other officers predicted the end of traditional tanks and armored personnel carriers. Tanks then in existence were too slow and heavy, while existing APCs did not provide enough protection against radiation. For example, in 1958 two officers argued that "The logistical requirements of the present heavy and medium tanks and the greatly increased range and penetrating power of small, direct-fire weapons will write finis to the fascinating career of these unwieldly giants." To them, the Army needed to develop an armored vehicle of no more than twenty tons that could move like a passenger vehicle and protect its crew from high levels of radiation. They speculated that it might be possible to create an electric field strong enough to protect the occupants from radiation.[164]

Contemporary officers argued that the pentomic organization was a "tremendous improvement… over the old triangular division."[165] Officers extolled the pentomic division as lean, powerful, and versatile. According to advocates, such units would be more easily able to disperse on the battlefield and hence would be less vulnerable. Enthusiasts argued that combat units would be capable of semi-independent operations over extended distances on a fluid battlefield for prolonged periods with minimal control or support from higher headquarters.[166] "Technological developments are occurring too rapidly for us to stand still or even to slow down," they argued. "We must not only keep abreast of these developments, but we must try to anticipate them if we are to build a combat force that will be victorious on the battlefield of the future."[167] The pentomic division also helped the Army compete with the Air Force and justified new weapons and additional personnel.

The pentomic organization was not without its critics, however. Some, such as the military analyst S. L. A. Marshall, argued that man remained central to the conduct of war and objected to the Army's emphasis on technology. Others were more pragmatic. They doubted whether the pentomic organization was tactically realistic, practical, and applicable.[168] Many questioned whether the Army would be capable of operating in a nuclear environment.

As it turned out, the pentomic division represented a dead end. The Army had neither the technology nor the money to implement it. Communication technology was not up to the task of allowing a commander to communicate with dispersed units on the nuclear battlefield. As General Paul Freeman, the former commander of the Continental Army Command, later recalled, "Every time I think of the... Pentomic Division, I shudder. Thank God we never had to go to war with it."[169] Even if the pentomic division had worked as advertised, it would have done nothing to prepare the Army for conventional conflicts or insurgencies. As a result, the Army began reorganizing its units in 1960.

The fifteen years that followed the end of World War II witnessed the most dramatic change in the size, organization, and technology of the U.S. armed forces in the twentieth century. The period saw the large-scale introduction of a wide variety of new weapons, such as jet aircraft, guided missiles, satellites, and integrated air defense systems. The services also fielded new organizations: The Air Force created missile units, the Navy launched nuclear-powered submarines, and the Army fielded pentomic divisions. Indeed, the nuclear age created whole new areas of military competition, including long-range nuclear strike, continental air defense, and the military use of space. It also led to a dramatic shift in the allocation of defense resources, with the Air Force taking the lion's share of the resources.

Perhaps more profoundly, the nuclear era led to a change in thinking about warfare. Civilian deterrence theorists such as Bernard Brodie, Albert Wohlstetter, and Thomas Schelling developed new theories about warfare, while systems analysts developed new techniques for acquiring complex weapon systems. By 1964, at least nine academic institutions offered systems engineering degrees. By 1970, there were an estimated eleven thousand systems engineers in the United States, most of whom were employed in aerospace.[170]

As dramatic as it was, the nuclear revolution failed to bear out the predictions of the more extreme technophiles. Nuclear weapons did not render war obsolete. Nor did ballistic missiles replace the Air Force's strategic bombers or the Navy's aircraft carriers. Indeed, the new elites that emerged from the nuclear revolution—such as missile officers in the Air Force and submariners in the Navy—remained subordinate to the traditional service elites.

Notes

1. Lawrence Freedman, *U.S. Intelligence and the Soviet Strategic Threat*, 2nd ed. (Princeton, NJ: Princeton University Press, 1986), 64.
2. Vannevar Bush, *Modern Arms and Free Men: A Discussion of the Role of Science in Preserving Democracy* (New York: Simon & Schuster, 1949), 123.
3. Robert Jackson, *High Cold War: Strategic Air Reconnaissance and the Electronic Intelligence War* (Sparkford, UK: Patrick Stephens, 1998), 38.
4. David Alan Rosenberg, "U.S. Nuclear War Planning, 1945–1960," in *Strategic Nuclear Targeting*, ed. Desmond Ball and Jeffrey Richelson (Ithaca, NY: Cornell University Press, 1986), 40.
5. R. Cargill Hall, "Postwar Strategic Reconnaissance and the Genesis of CORONA," in *Eye in the Sky: The Story of the Corona Spy Satellites*, ed. Dwayne A. Day, John M. Logsdon, and Brian Latell (Washington, DC: Smithsonian Institution Press, 1998), 105.
6. Jackson, *High Cold War*, 37.
7. William E. Burrows, *By Any Means Necessary: America's Secret Air War in the Cold War* (New York: Farrar, Straus & Giroux, 2001).
8. *GRAB: Galactic Radiation Background Experiment*, information paper (Washington, D.C.: Naval Research Laboratory, 2000) at http://code8200.nrl.navy.mil/grab .html (accessed June 23, 2006). See also Dwayne A. Day, "Ferrets Above: American Signals Intelligence Satellites During the 1960s," *International Journal of Intelligence and Counterintelligence* 17, no. 3 (2004): 449–67.
9. "POPPY Program Fact Sheet," at www.nro.gov/PressReleases/POPPY_Program_ Fact_Sheet.doc (accessed June 23, 2006).
10. Freedman, *U.S. Intelligence*, 65–66.
11. Ibid., 66.
12. Steven J. Zaloga, *The Kremlin's Nuclear Sword: The Rise and Fall of Russia's Strategic Nuclear Forces, 1945–2000* (Washington, DC: Smithsonian Books, 2002), 24.
13. Jeffrey T. Richelson, *The Wizards of Langley: Inside the CIA's Directorate of Science and Technology* (Boulder, CO: Westview Press, 2001), 13.
14. Ben R. Rich and Leo Janos, *Skunk Works* (Boston: Little, Brown, 1994), 130–31.
15. Gerald Haines, "The National Reconnaissance Office: Its Origins, Creation, and Early Years," in Day, Logsdon, and Latell, *Eye in the Sky*, 144.
16. Ibid., 145.
17. Freedman, *U.S. Intelligence*, 67.
18. Rich and Janos, *Skunk Works*, 151.
19. Ibid., 153, 157.
20. Ibid., 205, 211.

21. Ibid., 232.

22. Gene Poteat, "Stealth, Countermeasures, and ELINT, 1960–1975," *Studies in Intelligence* 42, no. 1 (1998): 51.

23. Gerald K. Haines and Robert E. Leggett, eds., *CIA's Analysis of the Soviet Union, 1947–1991* (Washington, DC: Center for the Study of Intelligence, 1993), 36.

24. Quoted in Ernest R. May, "Strategic Intelligence and U.S. Security: The Contributions of CORONA" in Day, Logsdon, and Latell, *Eye in the Sky*, 22.

25. Rosenberg, "U.S. Nuclear War Planning," 46.

26. Ibid., 47–48.

27. Maxwell D. Taylor, *The Uncertain Trumpet* (New York: Harper and Brothers, 1959), 131.

28. James M. Gavin, *War and Peace in the Space Age* (New York: Harper, 1958), 3–4.

29. Ibid., 243.

30. Ibid., 244.

31. Allan R. Millett and Peter Maslowski, *For the Common Defense: A Military History of the United States of America* (New York: Free Press, 1984), 513–14.

32. Richelson, *Wizards of Langley*, 32

33. Freedman, *U.S. Intelligence*, 72.

34. Day, Logsdon, and Latell, *Eye in the Sky*, 5.

35. Ibid., 39.

36. Kevin C. Ruffner, ed., *Corona: America's First Satellite Program* (Washington, DC: Center for the Study of Intelligence, 1995), xiv–xv.

37. Day, "Development and Improvement," 71.

38. Ibid., 7.

39. Central Intelligence Agency, *Intelligence Aspects of the Missile Gap*, TCS 11848/68, November 1968.

40. Freedman, *U.S. Intelligence*, 73.

41. See, for example, the translations in William R. Kintner and Harriet Fast Scott, *The Nuclear Revolution in Soviet Military Affairs* (Norman: University of Oklahoma Press, 1968).

42. Marc Trachtenberg, *History and Strategy* (Princeton, NJ: Princeton University Press, 1991), 4.

43. Rosenberg, "U.S. Nuclear War Planning," 38–39.

44. Millett and Maslowski, *For the Common Defense*, 476.

45. Rosenberg, "U.S. Nuclear War Planning," 38–39.

46. Ibid., 41–42.

47. Millett and Maslowski, *For the Common Defense*, 478.

48. Rosenberg, "U.S. Nuclear War Planning," 42, 48.

49. Millett and Maslowski, *For the Common Defense*, 517–18.

50. Robert Frank Futrell, *Ideas, Concepts, Doctrine: Basic Thinking in the United States Air Force, 1907–1960* (Maxwell AFB, AL: Air University, 1989), 428.

51. Rosenberg, "U.S. Nuclear War Planning," 44.

52. Millett and Maslowski, *For the Common Defense*, 512.

53. A. J. Bacevich, *The Pentomic Era: The U.S. Army Between Korea and Vietnam* (Washington, DC: National Defense University Press, 1986), 16.

54. George W. Baer, *One Hundred Years of Sea Power: The U.S. Navy, 1890–1990* (Stanford, CA: Stanford University Press, 1994), 350–51.

55. G. C. Reinhardt and W. R. Kintner, *Atomic Weapons in Land Combat* (Harrisburg, PA: Military Service Publishing, 1953), 9.

56. Gavin, *War and Peace*, 265.

57. Futrell, *Ideas*, 216.

58. Daniel Ford, "B-36: Bomber at the Crossroads," *Air and Space Magazine*, April/May 1996, 42–51.

59. Worden, *Rise of the Fighter Generals*, 32.

60. Futrell, *Ideas*, 252.

61. Ford, "B-36."

62. Zaloga, *The Kremlin's Nuclear Sword*, 18–19.

63. Ford, "B-36."

64. Carolyn C. James, "The Politics of Extravagance: The Aircraft Nuclear Program Project," *Naval War College Review* 53, no. 2 (Spring 2000): 158–90.

65. Ibid., 205–6.

66. Worden, *Rise of the Fighter Generals*, 106.

67. Millett and Maslowski, *For the Common Defense*, 515.

68. Futrell, *Ideas*, 479.

69. Robert Perry, *The Interaction of Technology and Doctrine in the USAF*, P-6281 (Santa Monica, CA: Rand Corporation, 1979), 9.

70. Kenneth P. Werrell, *The Evolution of the Cruise Missile* (Maxwell AFB, AL: Air University Press, 1996), 111.

71. Ibid., 111–12.

72. Ibid., 82.

73. Ibid., 85.

74. Ibid., 82.

75. Ibid., 85–86, 89.

76. Ibid., 93, 95.

77. Ibid., 96, 97.

78. Ibid., 100.

79. Jacob Neufeld, *Ballistic Missiles in the United States Air Force, 1945–1960* (Washington, DC: Office of Air Force History, 1990), 85.

80. Thomas P. Hughes, *Rescuing Prometheus* (New York: Pantheon Books, 1998).

81. Futrell, *Ideas*, 504.

82. Ibid., 504.

83. Werrell, *Evolution of the Cruise Missile*, 104.

84. Futrell, *Ideas*, 510.

85. Bush, *Modern Arms and Free Men*, 121.

86. Hughes, *Rescuing Prometheus*, 77.\

87. Perry, *Interaction of Technology*, 10.

88. Futrell, *Ideas*, 221.

89. Neufeld, *Ballistic Missiles*, 2.

90. Ibid., 71.

91. Ibid., 68.

92. Edmund Beard, *Developing the ICBM: A Study in Bureaucratic Politics* (New York: Columbia University Press, 1976).

93. Perry, *Interaction of Technology*, 11.

94. Hughes, *Rescuing Prometheus*, 104.

95. Ibid., 107–8.

96. Neufeld, *Ballistic Missiles*, 119.

97. Hughes, *Rescuing Prometheus*, 118.

98. Ibid., 70.

99. Neufeld, *Ballistic Missiles*, 179, 192.

100. Hughes, *Rescuing Prometheus*, 126.

101. Perry, *Interaction of Technology*, 14.

102. Neufeld, *Ballistic Missiles*, 121.

103. Ibid., 226.

104. Wilbur D. Jones Jr., *Arming the Eagle: A History of U.S. Weapons Acquisition Since 1775* (Ft. Belvoir, VA: Defense Systems Management Press, 1999), 352.

105. Hughes, *Rescuing Prometheus*, 51.

106. Ibid., 52, 54.

107. Ibid., 15.

108. Vincent Davis, *The Politics of Innovation: Patterns in Navy Cases* (Denver: University of Denver, 1967), 7.

109. Ibid., 11.

110. Ibid., 13.

111. Jerry Miller, *Nuclear Weapons and Aircraft Carriers: How the Bomb Saved Naval Aviation* (Washington, DC: Smithsonian Institution Press, 2001), 81.

112. Davis, *Politics of Innovation*, 16.

113. Ibid., 16–18.

114. William F. Trimble, *Attack from the Sea: A History of the U.S. Navy's Seaplane Striking Force* (Annapolis, MD: Naval Institute Press, 2005).

115. Werrell, *Evolution of the Cruise Missile*, 114.

116. Ibid., 116.

117. Ibid., 119.

118. Davis, *Politics of Innovation*, 33.

119. Ibid., 37–38.

120. Ibid., 39.

121. Baer, *One Hundred Years of Sea Power*, 353.

122. Davis, *Politics of Innovation*, 40.

123. Harvey M. Sapolsky, *Polaris System Development: Bureaucratic and Programmatic Success in Government* (Cambridge, MA: Harvard University Press, 1972).

124. Donald MacKenzie, *Inventing Accuracy: A Historical Sociology of Nuclear Missile Guidance* (Cambridge, MA: MIT Press, 1990), 145.

125. Michael Russell Rip and James M. Hasik, *The Precision Revolution: GPS and the Future of Aerial Warfare* (Annapolis, MD: Naval Institute Press, 2002), 63.

126. Owen R. Cote Jr., *The Third Battle: Innovation in the U.S. Navy's Silent Cold War Struggle with Soviet Submarines* (Newport, RI: Naval War College Press, 2002), 29.

127. Ibid., 21.

128. Ibid., 30.

129. Bacevich, *Pentomic Era*, 20.

130. Ibid., 19–20.

131. Ibid., 21.

132. John K. Mahon, "The Army's Changing Role," *Current History* (May 1954), 263.

133. Bacevich, *Pentomic Era*, 22, 53–54.

134. Theodore C. Mataxis and Seymour L. Goldberg, *Nuclear Tactics, Weapons, and Firepower in the Pentomic Division, Battle Group, and Company* (Harrisburg, PA: Military Service Publishing, 1958), vii.

135. Bacevich, *Pentomic Era*, 65.

136. Ibid., 56.

137. Marvin L. Worley Jr. *A Digest of New Developments in Army Weapons, Tactics, Organization, and Equipment* (Harrisburg, PA: Stackpole, 1959), 44.
138. Bacevich, *Pentomic Era*, 72.
139. Worley, *Digest of New Developments*, 17–18.
140. Bacevich, *Pentomic Era*, 82.
141. Ibid., 74.
142. James W. Bragg, *Development of the Corporal: The Embryo of the Army Missile Program* (Redstone Arsenal, AL: Army Ballistic Missile Agency, 1961).
143. Worley, *Digest of New Developments*, 30.
144. Mary T. Cagle, *History of the Sergeant Weapon System* (Redstone Arsenal, AL: U.S. Army Missile Command, 1971).
145. Ibid.
146. Mataxis and Goldberg, *Nuclear Tactics*, 39–40.
147. John W. Bullard, *History of the Redstone Missile System* (Redstone Arsenal, AL: U.S. Army Missile Command, 1965).
148. Bacevich, *Pentomic Era*, 87.
149. Michael H. Armacost, *The Politics of Weapons Innovation: The Thor-Jupiter Controversy* (New York: Columbia University Press, 1969).
150. Ibid., 88, 90.
151. Ibid., 100.
152. Ibid., 77–78.
153. Worley, *Digest of New Developments*, 38–39.
154. James R. Chiles, "Ring of Fire," *Air and Space* 20, no. 2 (June/July 2005), 34.
155. Ibid., 40–41.
156. Ibid., 323.
157. Bacevich, *Pentomic Era*, 68.
158. Ibid., 68–69.
159. Mataxis and Goldberg, *Nuclear Tactics*, 117; Robert A. Doughty, *The Evolution of U.S. Army Tactical Doctrine, 1946–76* (Ft. Leavenworth, KS: U.S. Army Command and General Staff Combat Studies Institute, 1979), 17.
160. Mataxis and Goldberg, *Nuclear Tactics*, 240.
161. Ibid., 241.
162. Worley, *Digest of New Developments*, 193–202.
163. Bacevich, *Pentomic Era*, 72–73.
164. Mataxis and Goldberg, *Nuclear Tactics*, 242, 240.
165. Worley, *Digest of New Developments*, 88.
166. Ibid., 89.
167. Mataxis and Goldberg, *Nuclear Tactics*, 100.
168. Bacevich, *Pentomic Era*, 131, 132.
169. Paul Freeman, *Sixty Years of Reorganizing for Combat: A Historical Trend Analysis* (Ft. Leavenworth, KS: Combat Studies Institute, U.S. Army Command and General Staff College, January 2000), 23.
170. Hughes, *Rescuing Prometheus*, 145.

2

Flexible Response, 1961–1975

THE DECADE AND a half following World War II was a highly disruptive period. The advent of nuclear weapons and long-range missiles triggered widespread change in the U.S. armed forces, producing both winners and losers. By contrast, the period from 1961 to 1975 represented in many ways a return to normalcy as the Kennedy administration's strategy of flexible response yielded a resurgence of the armed forces' traditional emphasis upon high-intensity conventional war.

The U.S. armed forces nonetheless faced new challenges. The combination of an active secretary of defense in the person of Robert S. McNamara and the widespread adoption of systems analysis as a technique for judging the merits of military programs tested the U.S. military. In cases involving systems both large and small, expensive and inexpensive, analysis by civilians in the Office of the Secretary of Defense (OSD) clashed with the long-standing preferences of the armed services, sparking interdepartmental disputes. The period also saw conflicts between civilian analysts and military officers over offensive nuclear forces and antiballistic missile (ABM) systems.

The Emergence of Systems Analysis

The need to reform the U.S. military was one of the main themes of John F. Kennedy's 1960 presidential campaign. Kennedy believed that the service bureaucracies, together with the Joint Chiefs of Staff, had become the major barrier to the efficient use of defense dollars.[1] To bring them under civilian control, Kennedy recruited Robert S. McNamara, the maverick president of the Ford Motor Company, to serve as secretary of defense. The primary tool McNamara introduced to impose rationality on the bureaucracy was systems analysis. Systems analysis offered a way to judge the relative merits of different weapon systems that could perform the same mission. McNamara was confident that systems analysis and other management techniques he had learned at Harvard Business School and implemented at Ford could also be applied to the Pentagon. He found kindred spirits at the Rand Corporation and brought many of its analysts, including Charles Hitch, Alain Enthoven, Henry Rowen, and Daniel Ellsberg, to Washington to lead a larger and more powerful OSD. The shift in civilian representation in the Defense Department was rapid: in 1960, there were 1,865 civilians in OSD, the Joint Staff, and other defense agencies. By mid-1962, their number had multiplied nearly twelvefold.[2]

Under McNamara, OSD took charge of how the services spent their money. McNamara demanded that the services adopt the Planning, Programming, and Budgeting System (PPBS), which had been popularized by business schools, as a way of planning the defense budget. His staff also reorganized the department's budget (and that of each of the services) into functional programs such as strategic forces, general-purpose forces, intelligence, and communications. This allowed the defense secretary and his staff to compare the relative value of competing programs that performed similar missions. It also forced the services to examine the financial implications of their choices of weapons.

McNamara created the office of the assistant secretary of defense for systems analysis to study defense programs. Whereas McNamara's predecessors had divided the overall defense budget then left it up to the services to figure out how to spend their share, McNamara distributed funds based on the relative effectiveness of a service's program. As Alain Enthoven and K. Wayne Smith, two of McNamara's aides, wrote, "as for the formulation of military needs, at the strategic level there is no such thing as a 'pure' military requirement, only alternatives with varying risks and costs attached. Choosing among these alternatives is the main job of the Secretary of Defense."[3]

The civilian systems analysts who populated OSD were highly confident of their ability to analyze strategic problems. As Enthoven and Smith wrote, "The problems of military strategy and force requirements, though complex, can be grasped, analyzed, and understood. They can be importantly, even if not wholly, quantified. Satisfactory answers can and should be found through a combination of judgment and analysis. Defense issues can and should be decided on their merits."[4]

The rise of the systems analysts alienated many in the military, who saw the civilians as arrogant and naive. As former Air Force Chief of Staff General Thomas D. White wrote in 1963, "In common with many other military men, active and retired, I am profoundly apprehensive of the pipe-smoking, tree-full-of-owls type of so called professional 'defense intellectuals' who have been brought into this nation's capital. I don't believe a lot of these over-confident, sometimes arrogant young professors, mathematicians and other theorists have sufficient worldliness or motivation to stand up to the kind of enemy we face."[5]

Similarly, retired General Curtis LeMay complained, "The military profession has been invaded by pundits who set themselves up as popular oracles on military strategy. These 'defense intellectuals' go unchallenged simply because the experienced professional active duty officers are officially prohibited from entering into public debate. The end result is that the military is often saddled with unprofessional strategies.... Today's armchair strategists, glibly writing about military matters to a public avid for military news, can do incalculable harm. 'Experts' in a field where they have no experience, they propose strategies based upon hopes and fears rather than upon facts and seasoned judgments."[6]

Nor were such sentiments the exclusive domain of the Air Force. As Vice Admiral Hyman Rickover testified, "The social scientists who have been making so-called cost effectiveness studies have little or no scientific training or technical expertise; they know little about naval operations.... Their studies are, in general, abstractions. They read more like the rules of a game of classroom logic than a prognosis of real events in the real world.... In my opinion we are unwise to put the fate of the United States in their inexperienced hands."[7]

The rise of systems analysis affected not only how the Defense Department bought new systems, but also what it bought. Systems analysis left its mark on a wide range of decisions, from the design of the F-111 aircraft to the decision whether to pursue ABM defenses.

Flexible Response

The incoming Kennedy administration was dissatisfied with its predecessor's strategy of massive retaliation, which held that the threat of a large-scale nuclear response was sufficient to deter both conventional and nuclear conflict. The new administration came into office determined to bolster the United States' conventional capabilities. One of the hallmarks of the Kennedy defense strategy was flexible response—the doctrine of meeting military threats symmetrically rather than automatically escalating to the use of nuclear weapons. To be credible, such a strategy required that the United States possess the capability to meet a Warsaw Pact attack with conventional force, at least initially. It also demanded the development of counterinsurgency capabilities in response to Soviet-backed "wars of national liberation" in the developing world.

Analysis conducted within McNamara's Pentagon suggested the feasibility of such a strategy. The analysis concluded that the size of the Soviet armed forces in Europe was half that estimated by the previous administration. For years, the military had told the Defense Department's civilian leadership that the Warsaw Pact, with 175 divisions, hopelessly outnumbered NATO, with only 21 divisions. However, OSD analysts argued that the number of divisions was a poor measure of the military balance, because the size and composition of Soviet divisions was much different than NATO divisions. In fact, they determined that the United States could field three times as many divisions if it organized along Soviet lines. An assessment of weapon stockpiles reinforced this view. In mid-1968, NATO had 55 percent of the tanks but as many artillery pieces and mortars, 150 percent of the antitank weapons and 130 percent of the armored personnel carriers of the Warsaw Pact.[8] As a result, the conventional defense of Western Europe—or at least a conventional opening phase of a war—seemed a real possibility.

To make flexible response a reality, the Kennedy administration increased the size of the U.S. armed forces by 250,000, the number of active Army divisions from eleven to sixteen, and tactical fighter wings from sixteen to twenty-one. Moreover, NATO, after American prodding, expanded to twenty-seven divisions, added five hundred aircraft to its inventory, and undertook a serious modernization program.[9]

Understanding the Soviet Military

Flexible response required a good understanding of Soviet military capabilities—not only the numbers of military formations and their equipment, but also detailed characteristics of weapon systems. The U.S. intelligence community developed a variety of techniques to understand the state of Soviet military research. The United States deployed sensors designed to identify the characteristics of new Soviet weapons on aircraft and satellites, under the sea, and on the periphery of the Soviet Union. In a number of cases, they deployed collectors clandestinely within the Soviet Union itself.[10] Photoreconnaissance aircraft and imagery satellites monitored Soviet test facilities. Signals intelligence (SIGINT) collectors eavesdropped on military exercises and administrative communications. Telemetry collectors intercepted and recorded the signals that weapons transmitted during tests.[11]

Another valuable technique for understanding Soviet military technology was the acquisition and analysis of foreign weapon systems. For example, the CIA obtained the guidance system of a SA-2 Guideline surface-to-air missile (SAM) that the Soviet Union had sold to Indonesia. Analysis of the weapon's capabilities allowed the Air Force to develop countermeasures to the missile for the B-52 bomber. The project also netted information on the SS-N-2 Styx antiship missile, Whiskey-class submarine, Komar-class guided-missile patrol boat, Riga-class destroyer, Sverdlov-class cruiser, Tu-16 Badger bomber, and AS-1 Kennel air-to-surface missile (ASM).[12]

The 1967 Arab-Israeli war was a boon to U.S. intelligence. It gave U.S. analysts not only an opportunity to observe a wide variety of Soviet weapons in combat, but also the ability to study a number of systems up close. This included the SA-2 SAM and its Fan Song radar, the AA-2 Atoll air-to-air missile (AAM), the SA-7 Strela SAM, and the guidance system for the Kennel and Styx missiles. In some cases, U.S. intelligence personnel were able to examine and evaluate weapons in the field; in other cases they analyzed systems in the laboratory.[13]

These and other efforts yielded confidence in the U.S. technological lead over the Soviets. As one 1967 National Intelligence Estimate (NIE) concluded, "We see no areas at present where Soviet technology is significantly ahead of that of the US." The authors nonetheless conceded that it was possible that the Soviets could move ahead of the United States. The Soviets possessed a massive military research and development program designed to prevent the United States from gaining a technological advantage or gaining one themselves. Analysts also noted that that although the U.S. intelligence community had the ability to detect new weapons during testing, they had much less insight into Soviet research and development. Although the intelligence community could make estimates concerning the next generation of major Soviet systems, analysts admitted that they could not forecast "the specific weapons which the Soviets will develop for introduction in the longer term, 10 or more years from now."[14]

The Army: A Return to Normalcy

The move to flexible response was accompanied by a reorganization of the U.S. Army. The service, which had in the 1950s undergone a radical transformation with the adoption of the pentomic division, abandoned that scheme and returned to more traditional organization and doctrine.

Beginning in the late 1950s, the Army sponsored a series of studies designed to find ways to overcome the shortcomings of the pentomic division. Instead of units optimized for the nuclear battlefield, the Army sought formations with sufficient firepower to cope with Soviet armored forces in conventional battle. The result was the Reorganization Objective Army Division (ROAD). In March 1961 Army Chief of Staff George H. Decker signed off on the ROAD scheme; two months later President Kennedy approved it for immediate implementation.[15]

Whereas the pentomic division embodied the Eisenhower administration's strategy of massive retaliation, ROAD was tailored for flexible response. Unlike the pentomic structure, ROAD units were designed primarily for combat on a conventional battlefield.[16] Each division had three combat brigade headquarters, to which battalions could be attached as needed. The adoption of ROAD, which remained the Army's standard organization until 1983, marked a return to the pre-pentomic divisional structure. In fact, it was an evolution of the World War II division structure. Its one significant novelty was the advent of mechanized infantry units: infantry mounted in tracked armored personnel carriers rather than trucks.[17]

In a larger sense, the adoption of ROAD marked the resurgence of traditionalists within the Army. Gone were the prophets of the nuclear battlefield. Instead, the Army returned to a more comfortable view of warfare, one shaped by the experience of World War II and Korea.

Tank Modernization

The 1960s saw the Army undertake a large-scale modernization program as the service sought weapons to implement flexible response. In 1960, it accepted into service the M-60 tank. The M-60, an evolutionary development of the M-48 Patton tank, was equipped with a 105mm main gun. The main difference between the M-60 and its predecessor was the fact that it had a welded hull, rather than a cast hull like the M-48. Two years later, the Army began deploying the M-60A1 with a new turret, increased armor protection, and a new ammunition storage system.

The next variant of the M-60, the M-60A2, was an attempt to field a radically different type of tank. First accepted in 1974, it featured a new turret fitted with a combined gun and missile launcher for the MGM-51 Shillelagh antitank guided missile (ATGM).

The development of the Shillelagh, the most complex system the Army had fielded to date, was fraught with problems. The contractor that developed the missile significantly underestimated the complexity and magnitude of the task of developing the weapon. The missile's propellant, igniter, tracker, and infrared command link all experienced problems. Moreover, because ATGMs were relatively new, solving these problems proved to be extremely difficult.[18]

The missile had operational shortfalls as well. One problem was that it was a line-of-sight weapon that required the gunner to keep the target in the crosshairs throughout the missile's flight. This was a challenge, particularly in Europe, where trees, smoke, rain, hills, and darkness could block the gunner's view of the target.[19]

The M60A2 had its share of problems as well. The tank, sometimes derisively known as the "starship", featured a complex fire-control system. Moreover, its high profile and limited cross-country mobility limited its utility. As a result, only 543 were produced; first adopted in 1975, it was phased out six years later.

If the M-60A2 proved to be a misstep, the next evolution of the M-60, the M-60A3, was much more successful. First accepted in 1978, the tank improved upon the basic design of the M-60A1 by adding a more advanced fire-control system that included a laser rangefinder, solid-state ballistic computer, and crosswind sensor. It also mounted a thermal sight to allow its crew to spot enemy vehicles in inclement weather. Less radical than the M60A2, the M60A3 in the end proved more successful. Indeed, Marine units employed the tank successfully in the 1991 Gulf War.

Another technologically ambitious project was the tank that was to have been the successor to the M60 family, the joint U.S.–West German MBT-70 project. Initiated in August 1963, the tank was designed to be an armored vehicle for the computer age. It was to have the capability to fire either Shillelagh or conventional ammunition and would feature a stabilized fire-control system and laser rangefinder.[20] As the project moved from development to production, however, it became clear that the United States and Germany had different design philosophies and technical requirements. As a result, the cost of the system rose and its schedule stretched. In January 1970, the program was terminated. The follow-on to the tank, the XM803, survived for two years before being terminated in January 1972. Its successor, the XM815, would eventually be fielded as the M1 Abrams.

The Air Force: The Rise of Tactical Air Power

The advent of flexible response led the Air Force to place greater emphasis upon conventional air power. Moreover, it began to de-emphasize long-range bombers in favor of theater aircraft, a trend that had implications for not only the structure of the Air Force but also its culture. Indeed, during the 1970s as a result of the Vietnam War, the fighter community came to dominate the leadership of the U.S. Air Force.

The F-111

The development of the TFX (Tactical Fighter Experimental), which was fielded as the F-111 Aardvark, was a case study in the clash between the military services and civilian systems analysts. The TFX grew out of two different aircraft programs. First, the Air Force's Tactical Air Command sought a replacement for the F-105 Thunderchief fighter-bomber. The Air Force wanted an aircraft that could carry nuclear weapons internally, ferry across the Atlantic without refueling, operate from semi-prepared airfields in Europe, and fly at Mach 2.5 at high altitude and subsonic at low altitude. Second, the Navy sought a carrier-based, fleet-defense strike fighter.[21]

On February 16, 1961, McNamara directed the Navy and Air Force to design a single aircraft fighter-bomber to meet both sets of requirements. This, however, was easier said than done. The Navy and Air Force requirements grew out of different missions, operational environments, and even notions of air power. The Air Force wanted an aircraft with a large payload and efficient operation at high and low altitude. The Navy wanted a lighter, more maneuverable aircraft capable of taking off from and landing on an aircraft carrier.[22] Although the services felt it unrealistic to combine the two requirements, they did their best to comply with McNamara's request.

The process of choosing the manufacturer of the TFX fueled controversy. Although Boeing won all four stages of the competition, in November 1962 McNamara chose rival General Dynamics, which promised to build the TFX in Vice President Lyndon B. Johnson's home state of Texas.[23] Grumman, based in New York, was selected to build the Navy version of the aircraft. The fact that both Texas and New York were swing states in the 1964 election fueled suspicion that partisan politics had influenced the selection of General Dynamics and Grumman. Although the award was based upon cost effectiveness and efficiency, the decision irritated the Air Force chief of staff and the Navy chief of naval operations, both of whom wanted Boeing to build the aircraft. In 1963, a special Senate subcommittee chaired by Senator John L. McClellan of Arkansas held hearings on the contract award, and Secretary of the Navy Fred Korth, who had links to General Dynamics in his home state of Texas, resigned in November.

Although the TFX was supposed to meet both Air Force and Navy requirements, in practice it was designed first and foremost to Air Force specifications and then modified as much as possible to meet the constraints of a carrier-based aircraft. The first Air Force version, the F-111A, flew in December 1964; the first Navy F-111B flew the following May. Although they demonstrated the feasibility of a variable sweep wing to allow both high-speed flight and long range, the aircraft was sluggish and underpowered. Moreover, the F-111A's afterburning turbofan engines stalled and surged, a problem that led to numerous changes in their design. Although the problems were eventually solved, they triggered major cost overruns. For low-altitude penetration of defended areas, the F-111 featured an automatic terrain-following radar that allowed it to fly as low as two hundred feet at high speeds. Overall, however, the aircraft was too complicated, too big, and too expensive.[24]

The Navy's F-111B was designed to defend carrier battle groups against Soviet bombers and missiles. However, the aircraft was poorly suited to operations aboard aircraft carriers: It was too heavy and too long to fit on carrier elevators. In 1968, the Navy TFX was cancelled.

The Air Force was left with a design that had been compromised by the Navy's requirements. The F-111A was rushed into combat in Southeast Asia in 1968. However, the aircraft's combat debut was plagued by a number of problems, including a lack of trained, experienced crews. As a result, it accumulated an undistinguished record.[25]

It was not until the 1980s that the F-111 was to demonstrate fully its worth. Aircraft from the 48th Tactical Fighter Wing at Lakenheath in the United Kingdom played a key role in the U.S. strike on Libya on April 15, 1986. During the operation, aircraft equipped with Pave Tack laser-designation pods delivered 2,000-pound GBU-10 and 500-lb GBU-12 laser-guided bombs against Libyan targets.[26] During the 1991 Gulf War, the F-111 was only one of two Air Force aircraft capable of delivering laser-guided bombs.

The F-4 Phantom II

A more successful case of developing an aircraft to meet the needs of more than one service was the F-4 Phantom II. In mid-1954, the Navy's Bureau of Aeronautics issued a request for a new all-weather carrier-based fleet defense interceptor. McDonnell Douglas submitted proposals for both a single-engine fighter, designated F3H-E, and a twin-engine design, the F3H-G. The Navy chose the twin-engine model. Although McDonnell's mock up included four 20mm cannon, the Navy wanted the fighter to be equipped only with Sidewinder and Sparrow AAMs. The first XF4H-1 was rolled out on May 8, 1958, and took its initial flight later in the month. Carrier trials began in fall 1959.

McNamara saw the acquisition of the F-4 as another opportunity to achieve commonality, and in 1961 he requested that the Air Force evaluate the aircraft. Although commonality didn't often work, the F-4 was an exception: eventually the Air Force acquired twice as many of the aircraft as the Navy and Marine Corps combined.

Although the aircraft proved to be a success, the services soon regretted the decision to forgo arming the Phantom with a cannon. F-4s frequently found themselves engaging North Vietnamese fighters in dogfights at ranges too close to use their Sidewinders, which in any event often proved unreliable. As a result, both Navy and Air Force F-4s were fitted with cannons. Initially the aircraft were outfitted with the SUU-16/A external gun pod. However, the F-4E, which performed its initial flight test in June 1967, was equipped with a M61A1 six-barrel Gatling-type cannon on the underside of its nose.

The Navy: Antisubmarine Warfare

During this period, the U.S. Navy harnessed America's technological advantage to compete with the Soviet Union in undersea warfare. The U.S. Navy mastered the technology for detecting Soviet submarines while also hiding U.S. submarines from the Soviets.

Although the Soviet surface fleet did not pose a major threat to the United States and its allies in the 1960s and early 1970s, the growth of the Soviet submarine force was cause for concern. In 1955, the Soviet navy conducted its first launch of a ballistic missile from a submerged submarine. The first Soviet submarine to be deployed with ballistic missiles was the diesel-powered Project 629 (Golf). The submarine, which was produced between 1958 and 1962, carried 3 R-13 (SS-N-4 Sark) missiles in its sail.[27] However, the combination of the Golf and SS-N-4 had significant limitations. The Golf's diesel propulsion limited the length of its patrols, while the SS-N-4's 600 km range meant that the Golf would have to approach the shores of the United States before launching an attack. To make matters worse, the submarine had to surface to launch its missiles. Finally, liquid-fueled sea-launched ballistic missiles (SLBMs) were dangerous, with fuel so corrosive that submarines were rarely loaded with live weapons.[28]

The late 1950s saw the deployment of the first generation of Soviet nuclear submarines: the Project 658 (Hotel) SSBN, Project 659 (Echo) SSGN, and Project 627 (November) SSN. These submarines' nuclear propulsion allowed them to stay on station for a longer period of time, remedying one of the deficiencies of diesel-powered boats. They were, however, noisy, and experienced mechanical problems. The lead boat in the Hotel class, K-19, suffered a number of accidents, culminating in a severe fire on February 24, 1972, that cost the lives of twenty-eight crewmembers.[29]

The emergence of the Soviet submarine threat led the U.S. Navy to develop new approaches to antisubmarine warfare (ASW). Whereas ASW operations in World War II had used active sonar, the Navy's new approach to ASW was based upon the development of passive sonar, particularly Low Frequency Analysis and Ranging (LOFAR), a technique that offered a particularly powerful tool for detecting submarines at a long distance.[30] The Navy sponsored research on LOFAR at Bell Laboratories under the cover name Project Jezebel.[31] Passive acoustics played an important role in ocean surveillance, submarine vs. submarine operations, and maritime patrol operations.

The Sound Surveillance System (SOSUS) network, which employed passive acoustics and LOFAR processing, revolutionized ocean surveillance. The network consisted of arrays of hydrophones spaced along undersea cables emplaced on the sea floor. Developed by scientists at Bell Labs and Columbia University's Hudson Lab, the Navy installed the first test array off Eleuthera in the Bahamas in 1951. The following year, the Navy decided to emplace arrays across the eastern seaboard; two years later, it began deploying arrays off the Pacific coast and Hawaii. These installations were completed in 1958. The following year, the Navy installed an array off Argentia, Newfoundland.[32]

SOSUS arrays came ashore at Naval Facilities, or NavFacs, where computers processed and sailors analyzed the signals. The Navy also established evaluation centers, known as Ocean Surveillance Intelligence System (OSIS) nodes, where sailors assessed processed acoustic information and combined it with other data, such as high-frequency direction-finding information.[33] A crucial adjunct to the emergence of SOSUS was the development of acoustic intelligence as a distinct discipline within the naval intelligence community. The community not only developed the ability to locate and track Soviet ships and submarines, but to identify individual ships and submarines by their acoustic "fingerprint."[34]

SOSUS gave the U.S. Navy the ability to detect and classify Soviet submarines at astounding ranges. The first-generation Soviet nuclear submarines were extremely vulnerable to passive acoustics because their propellers were loud and rotating machinery mounted to the submarine's hull created signatures loud enough for the United States to detect and track at long distances. In July 1962, in the Navy's first detection of a Soviet nuclear submarine, sailors monitoring the SOSUS array off Barbados monitored a Soviet nuclear submarine as it crossed the Greenland–Iceland–United Kingdom gap.[35]

The development of SOSUS alerted submarine designers to the fact that Soviet passive acoustics could detect U.S. submarines. As a result, they sought to design boats that could not be detected by SOSUS. Improvements in quieting, sonar performance, sonar system integration, and tactical and operational analysis led to gains in quieting in successive generations of U.S. submarines.[36]

The deployment of the Sturgeon (SSN-637) class gave the Navy an enhanced ability to track Soviet submarines covertly. The quiet submarine's wide-angle bow sonar allowed it to track Soviet submarines. As Owen Coté has written, "The 637s made it feasible to develop tactics for routine, covert tracking operations that could be implemented on a force-wide basis."[37] The Navy launched thirty-seven of the submarines, armed with twenty-four torpedoes and antiship missiles, between 1966 and 1975.

In 1968, the Soviets began deploying their second generation of nuclear submarines, the Project 670 Skat (Charlie) SSGN, Project 671 Kefal (Victor) SSN, and Project 667A Navaga (Yankee) SSBN. The deployment of the Yankee, armed with sixteen 2,400 km R-27 (SS-N-6 Serb) SLBMs, gave the Soviets greater flexibility in their nuclear force. The submarine's designers attempted to reduce its noise signature by equipping it with quiet propellers, lining its inner hull with sound-absorbing material, and coating its outer hull with soundproof rubber.[38] The submarine entered service in 1967, and the following year the Soviets began sending Yankees on patrol off the east coast of the United States. Three years later they showed up off the west coast.[39]

The deployment of the follow-on to the Yankee, the Project 667B (Delta I) SSBN beginning in 1972 caused greater concern. Armed with twelve

R-29 (SS-N-8 Sawfly), the first Soviet SLBM with truly intercontinental reach, the submarine gave the Soviets the ability to hold American targets at risk from the seas adjacent to and adjoining the Soviet Union. As a result, Soviet SSBNs no longer had to pass through the U.S. SOSUS barrier. Rather, they could be protected in so-called bastions near the Soviet coast. The Soviets began adopting such a strategy in the late 1970s, when the Soviets learned, through the Walker espionage ring, just how vulnerable their SSBNs were to detection by the U.S. Navy.[40] In all, the Soviets deployed a total of forty-two Delta-class SSBNs in four classes.[41]

The Navy's ultimate fear—a truly quiet Soviet nuclear submarine—took much longer to emerge than had been feared. The Victor III SSN, first deployed in the late 1970s, was the first submarine to surprise U.S. analysts with its quietness. It was followed by the Akula SSN, which approached parity with the United States in quieting. The Soviet gains were the result of applying a variety of techniques to reduce the noise a submarine generated, as well as the acquisition of advanced manufacturing machinery from Japan that allowed the Soviets to manufacture more advanced propellers. As a result, in the mid-1980s the United States faced a Soviet submarine that could elude tracking by SOSUS and tactical ASW platforms using passive acoustics.[42]

The Nuclear Balance

Although the Kennedy administration inaugurated a conventional buildup, it did not ignore the U.S. nuclear posture. Indeed, conventional modernization was meant to make the nuclear deterrent more credible. Both Kennedy and his advisors were, however, skeptical of the doctrine of massive retaliation and its embodiment in Single Integrated Operational Plan (SIOP)-62. In February 1961, Secretary of Defense McNamara sent Kennedy a letter in which he identified the vulnerability of U.S. forces to attack and the lack of flexible response options as concerns.[43] Moreover, although SIOP-62 emphasized military targets, it also put a lot of civilians at risk. Maxwell Taylor, the President's Military Representative, characterized SIOP-62 as "a rigid, all-purpose plan, designed for execution in existing form, regardless of circumstances.... SIOP-62 is a blunt instrument."[44]

As a result, Kennedy ordered a complete review of the SIOP. In so doing, the administration faced a number of alternatives. During his first week in office, McNamara received a briefing from RAND analyst William Kaufmann recommending a counterforce nuclear strategy.[45] According to such an approach, the United States would seek the capability to destroy all or most Soviet nuclear forces while still retaining an assured ability to threaten Soviet cities. The Air Force embraced counterforce, not least because it justified the service's nuclear programs. The Navy, by contrast, favored a counter-city strategy of minimum deterrence, one well suited to the limited accuracy of the Polaris.

A third option was to adopt a controlled nuclear response. However, the Joint Chiefs of Staff were skeptical that such a strategy was feasible. As Army General Lyman Lemnitzer, the chairman of the Joint Chiefs of Staff, put it, "My personal judgment is that we do not now have adequate defenses, nor are our nuclear retaliatory forces sufficiently invulnerable, to permit us to risk withholding a substantial part of our effort, once a major thermonuclear attack has been initiated."[46]

The SIOP that resulted from the review included five categories of targets: Soviet strategic nuclear forces; other elements of the Soviet military located away from cities, such as air defenses along bomber routes; military forces near cities; command and control facilities; and an all-out urban attack.[47]

In a commencement address at the University of Michigan on June 16, 1962, McNamara announced that the "principal military objectives, in the event of a nuclear war stemming from a major attack on the Alliance, should be the destruction of the enemy's military forces, not of his civilian population.... We are giving a possible opponent the strongest imaginable incentive to refrain from striking our own cities."[48] Such a strategy was possible both because of U.S. nuclear strategy and improved intelligence regarding the Soviet strategic posture provided by reconnaissance satellites.

McNamara soon became disenchanted with counterforce, as much for budgetary reasons as strategic logic. He realized that counterforce offered no logical limit to the size of the U.S. nuclear arsenal. In December 1963, McNamara gave President Johnson a new Draft Presidential Memorandum that emphasized deterrence over counterforce as the main mission of U.S. nuclear forces. The new requirement was for the United States to retain sufficient forces to survive a surprise Soviet first strike and destroy the Soviet government and military leadership as well as a large portion of the Soviet population and industrial base.[49]

Between 1964 and 1966, U.S. declaratory policy featured both the need to inflict unacceptable damage on the Soviet Union ("assured destruction") as well as to limit the ability of the Soviet Union to inflict damage on the United States ("damage limitation"). Over time, McNamara continued to play down damage limitation, so that by 1967 his rhetoric focused on assured destruction.[50] As he put it in February 1965, "A vital first objective, to be met in full by our strategic nuclear forces, is the capability for Assured

Destruction. What kinds and amounts of destruction we would have to be able to inflict in order to provide this capability cannot be answered precisely. But, it seems reasonable to assume the destruction of, say, one-quarter to one-third of its population and about two-thirds of its industrial capacity... would certainly represent intolerable punishment to any industrialized nation and thus should serve as an effective deterrent."[51]

The shift in U.S. declaratory policy to assured destruction was based upon a judgment as to what amounted to "diminishing returns" and what would deter a "rational actor." It was not based on what would deter the Soviets. Indeed, there was exceedingly little analysis of the Soviets as competitors.

In January 1974, Secretary of Defense James Schlesinger initiated a major reorientation of U.S. nuclear policy. Stating "that the destruction of enemy cities 'should not be the only option and possibly not the primary option' of the United States in the event of war," Schlesinger called for the establishment of a series of limited nuclear options.[52] Schlesinger also wanted to be able to suppress the Soviet Union's economic recovery by holding at risk 70 percent of Soviet industry. SIOP-5, prepared under this guidance, was approved in December 1975 and took effect on January 1, 1976.[53]

The late 1960s and early 1970s led to a questioning not only of U.S. nuclear doctrine, but also of the overall approach to the U.S. Soviet competition. In a classified report published in 1972 Andrew W. Marshall, then an analyst at the Rand Corporation, questioned the assumption that the United States was engaged in an action-reaction arms race. As he wrote, "Commonly used hypotheses about the nature of the strategic arms race, or about the U.S.-Soviet interaction process (claiming a closely coupled joint evolution of U.S. and Soviet force postures), are either demonstrably false or highly suspect. The more serious classified studies of the interaction process almost uniformly present a picture of a much more complex, slower moving action-interaction process than that asserted by arms control advocates."[54]

Marshall also offered a critique of the way the United States was conducting its competition with the Soviet Union. He noted that the United States had followed a rich nation's strategy of attempting to compete with the Soviet Union in all areas of technology. However, the fact that the United States would have relatively fixed resources to devote to defense while the Soviets were able to apply concerted effort to develop advanced technology was making such a strategy untenable. As he put it, "The Soviets are closing the military R&D gap, probably one of their top priorities since World War II. Previously the United States could support a policy of staying ahead in *all* of the areas of technology it most cared about. The list has to be smaller now, and the United States may need a new R&D strategy."[55]

Marshall argued that the United States needed to develop a strategy for competing with the Soviets over the long term. Specifically, he recommended that the Untied States needed to move the competition into areas where the United States had an enduring advantage or the Soviets a specific disadvantage. As he noted, "To some extent the United States can probably force increased expenditures on the Soviet Union in specific areas, thus preventing their fixed resources from being spent on other things that may be more threatening to the United States."[56]

Shortly after publishing the monograph, Marshall was brought to Washington, D.C., to work for National Security Advisor Henry A. Kissinger. In October 1973, Secretary of Defense Schlesinger appointed him the Defense Department's first Director of Net Assessment. In that position he advised the Secretary of Defense on long-term strategy issues.

The Soviet Nuclear Arsenal

The Soviet missile force grew slowly in the early 1960s. The Soviets possessed 16 ICBMs in 1961, 56 in 1962, 122 in 1963, and 189 in 1964. On November 1, 1961, the Strategic Rocket Forces declared four regiments of the R-16 (SS-7 Saddler) ICBM operational. Although the SS-7 was the first Soviet missile that did not need to be fueled immediately prior to launch, it was deployed on soft launch pads and took up to three hours to ready. The R-9 (SS-8 Sasin) suffered a troubled development and was not accepted until July 1965, four years behind schedule.[57]

The state of the Soviet nuclear arsenal was a key consideration in the development of U.S. nuclear forces. The technical characteristics of these missiles had strategic implications, including whether the Soviets could destroy U.S. ICBMs in a preemptive strike or overwhelm U.S. missile defenses. Because the Soviet Union was a denied area, the U.S. intelligence community was faced with the need to develop innovative approaches to collect and analyze intelligence. U.S. intelligence agencies collected optical, radar, and telemetry data on Soviet missile tests to determine the size and shape of their reentry vehicles and therefore to estimate the yield of their warheads. They used a chain of eavesdropping stations to intercept telemetry signals. And the CIA used a 150-foot dish antenna at Stanford University to monitor signals of Soviet radars as they bounced off the moon.[58]

Beginning in January 1960, the United States embarked upon an expansive effort to collect information on Soviet ballistic missile tests into the Pacific. Various units gathered electronic and other forms of intelligence on Soviet ICBMs before, during, and after their flight. U.S. submarines collected a unique portion of this data. No other platform was able to observe the detachment of the data capsule from the missile nose cone, its descent by parachute, and recovery by Soviet missile range instrumentation ships.[59]

The early Brezhnev years saw the Soviet Union drive for parity with the United States with the testing and deployment of their second-generation ICBMs: the R-36 (SS-9 Scarp), UR-100 (SS-11 Sego), and RT-2P (SS-13 Savage). However, the U.S. intelligence community consistently underestimated the scope and pace of the Soviet buildup. The 1964 NIE on Soviet strategic forces argued that Russian aspirations might not extend beyond four hundred to five hundred missiles, and certainly would not exceed seven hundred. It concluded, "We do not believe that the USSR aims at matching the US in numbers of intercontinental delivery vehicles. Recognition that the US would detect and match or overmatch such an effort, together with economic constraints, appears to have ruled out this option."[60] In fact, by 1970 the Soviets had deployed 1,158 ICBMs.[61]

The defense department and the intelligence community tended to extrapolate past Soviet behavior into the future rather than consider the possibility of discontinuity. They expected the Soviet buildup to proceed in an orderly, deliberate fashion.[62] In so doing, they failed to consider the possibility that the Soviets might have objectives distinct from those of the United States. As one intelligence community postmortem concluded, analysts overestimated barriers to matching the United States, overestimated Soviet concern about provoking new U.S. deployments, and underestimated the impetus for an expansion of the Soviet nuclear force because analysts doubted that anything past equality would be of military value.[63]

McNamara shared these assumptions regarding the Soviet Union. As he put it in April 1965, "They do have the ability to catch up [with the United States in strategic forces] by 1970. Therefore, I cannot give you any final estimate what their 1970 force will be, because next year they could change their plans. But I can say that their rate of expansion today is not such as to allow them even to equal, much less exceed, our own 1970 force... [This] means that the Soviets have decided that they have lost the quantitative race, and they are not seeking to engage us in that contest. It means there is no indication that the Soviets are seeking to develop a strategic nuclear force as large as ours."[64]

The increase in Soviet ICBM deployments raised the possibility of a counterforce threat. The SS-9's ability to carry a large payload aroused the most controversy. One faction argued that the missiles were designed to strike U.S. cities. Another argued that the missiles could be equipped with multiple independently targeted reentry vehicles (MIRVs) to neutralize the U.S. Minuteman force.[65] Although silos would provide protection against inaccurate missiles, they would be ineffective against accurate ones. One particular concern was that the Soviets would use MIRVed SS-9s to strike Minuteman launch control centers.[66]

Determining the number of reentry vehicles the SS-9 would carry, the yield of its warheads, and their accuracy was critical to understanding the missile's mission. The United States carefully monitored the missile's performance as soon as it entered testing in 1963.[67] Concern grew in August 1968, when the Soviets tested a multiple reentry-vehicle (MRV) system for the missile. The R-36P (SS-9 Mod 4 "Triplet") became operational in 1971. To destroy the entire Minuteman force, the Soviets would have required 420 of the MIRVed SS-9s, each with three five-megaton warheads, reasonable accuracy, and the ability to retarget their missiles.[68] Whether the Soviets were in fact trying to acquire such a capability was a point of contention within the U.S. government. The argument over the missile remained unsettled until the Soviets finally began testing their next generation of ICBMs in 1973—missiles that clearly possessed greater accuracy and the ability to hit multiple, independent targets and thus posed a threat to U.S. ICBMs.[69]

U.S. Nuclear Forces

John F. Kennedy campaigned for the presidency on the need to strengthen America's nuclear posture and to redress the supposed missile gap between the United States and Soviet Union. He was particularly concerned about the vulnerability of U.S. nuclear forces. To reduce that vulnerability, Kennedy accelerated the deployment of the solid-fuel Minuteman ICBM, which could be launched more quickly than the liquid-fueled Atlas and Titan missiles. He also sped up the Polaris SLBM program to provide an invulnerable deterrent force while also retiring the first generation of nuclear missiles, the Snark, Thor, Jupiter, and Regulus. He also improved nuclear command and control by establishing alternative national command centers, improving strategic communications, and establishing safeguards against accidental nuclear launch.

The move to missiles reduced the importance of manned aircraft in the U.S. nuclear force. The Pentagon decided not to purchase additional B-52s and to phase out the B-47. It also cancelled the Skybolt air-launched ballistic missile and XB-70 Valkyrie bomber programs.[70] The cancellation of Skybolt and the RB-70 consolidated the dominance of ballistic missiles within the U.S. nuclear posture.

When Kennedy came to office, the U.S. ICBM force consisted of nine Atlas missiles, deployed aboveground and unprotected in two clusters. Two well-placed nuclear detonations could have completely destroyed them. To redress this vulnerability, Kennedy accelerated the acquisition of the Minuteman ICBM, a missile that was less costly than existing systems, easier to deploy in hardened silos, and easier to maintain on alert.[71]

The Air Force originally envisioned deploying the missiles in silos. Motivated in part by Navy criticism of ICBM vulnerability, however, it also investigated fielding at least a portion of the force on railroad cars. Although the Air Force's Strategic Air Command (SAC) led the campaign to make the Minuteman mobile, studies revealed significant operational and logistical problems with the scheme.[72] Mobile basing was also expensive—costing up to ten times as much per missile. Finally, the deployment of SSBNs beginning in 1960 gave the United States an invulnerable deterrent. As a result, the Air Force cancelled the rail-basing plan in 1962.[73]

Analysis confirmed the benefits of deploying the U.S. ICBM force in hardened silos. An assessment conducted by Albert Wohlstetter of the Rand Corporation showed that the U.S. inventory of 120 ICBMs could be destroyed by a small Soviet missile force if unsheltered. However, he calculated that it would take more than 7,600 Russian missiles to destroy eighty percent of the force if it were housed in silos hardened to resist 200 psi of overpressure.[74]

Accordingly, Minuteman I was housed in hardened, widely dispersed underground silos. An underground launch control center monitored each flight of ten launch facilities, with five flights per squadron. The HSM-80A Minuteman IA achieved its initial operational capability in December 1962 with twenty missiles; by the end of February 1963 a full squadron was on alert. A total of 150 of the missiles were deployed. The first HSM-80B Minuteman IB, equipped with an improved second-stage motor, new reentry vehicle, and larger warhead, entered service in September 1963. The Air Force deployed 650 by June 1965.

The following year the Air Force initiated the Minuteman Force Modernization Program to replace all Minuteman I missiles with either Minuteman II or Minuteman III models. The LGM-30F Minuteman II had an improved second stage, a dramatically improved guidance system, and was equipped with solid-state circuitry. In fact, the Minuteman II's guidance and control system was three times more accurate than the Minuteman I's, a system that had been developed only four years before. This accuracy gave Minuteman II the ability to strike hardened targets, such as Soviet missile silos and command centers.[75] First operational in October 1965, the Air Force deployed 1,000 Minuteman IIs in all.

The LGM-30G Minuteman III was the world's first ICBM to carry MIRVed warheads. It featured an enlarged third stage as well as a new warhead section, or "bus," that housed the guidance system, its own liquid-fueled rocket motor, and three Mk 12 MIRVs, each with a 170-kiloton W-62 thermonuclear warhead. The missile went into regular development in 1966 and was first deployed in April 1970. A total of five hundred Minuteman IIIs were fielded.

Protecting the U.S. Nuclear Force

The growth of the Soviet nuclear arsenal led not only to efforts to improve the U.S. nuclear force, but also to ensure its survivability. One option was to improve passive defense measures such as warning of an impending attack and hardening missile silos. In the early 1950s the United States began deploying the Distant Early Warning (DEW) and Pine Tree lines of radar installations across northern and central Canada to warn of Soviet aircraft approaching North America over the North Pole. In late 1959, the Defense Department's Advanced Research Projects Agency (ARPA) opened the 474L System Program Office, which was tasked with developing techniques and equipment for tracking space objects and detecting Soviet ICBMs. By the mid-1960s, it had activated three Ballistic Missile Early Warning System (BMEWS) radars at Thule Air Base, Greenland; Clear Air Force Station, Alaska; and RAF Fylingdales Moor, England. BMEWS could provide the capability to detect an incoming ICBM attack at a range of approximately 3,000 nautical miles and provide fifteen minutes of warning, giving the president of the United States the ability to order the launch of U.S. missiles before the Soviet missiles reached their targets.

ABM

Another approach was to defend against a Soviet missile attack. One of the most important questions in the development of the U.S. nuclear posture in the 1960s and early 1970s was whether the United States should deploy ABM defenses. The case in favor of ABM was fairly straightforward: politically, strategically, and morally, it made sense to protect the United States against the threat of Soviet ballistic missiles. ABM advocates argued that even a marginally effective system could be militarily useful and save American lives.

Opponents of ABM advanced two seemingly contradictory arguments. The first was that an effective defense against ballistic missiles was unachievable. Even if only a handful of Soviet missiles would leak through a U.S. defense, they agued, the result would be devastating. Moreover, elements of the defensive system, particularly its radars, would themselves be vulnerable to attack. The second argument was that attempting to field an ABM system would itself be destabilizing. As Jerome Wiesner and Herbert York wrote, "Paradoxically, one of the potential destabilizing elements in the present nuclear standoff is the possibility that one of the rival powers might develop a successful antimissile defense. Such a system, truly airtight and in the exclusive possession of one of the powers, would effectively nullify the deterrent forces of the other, exposing the latter to a first attack against which it could not retaliate."[76]

McNamara was skeptical of ABM, believing that a truly effective system was not feasible and that the attempt to acquire one would spark an arms race.[77] However, Soviet ABM development, U.S. technological breakthroughs, and support for ABM among the U.S. military conspired to weaken his opposition.

In 1961, the U.S. intelligence community detected the construction of installations analysts suspected to be part of an ABM effort near Leningrad. The following year, the Soviets deployed some thirty SA-1 Guild SAMs, thought to possess a limited ballistic missile defense (BMD) capability, around the city. The Soviets then began deploying SA-5s along the so-called Tallinn Line, which stretched in a wide arc from Archangelsk to Riga along the approach corridor for U.S. ICBMs. The main question facing the intelligence community was whether the system was designed for defense against U.S. bombers or missiles.[78] The consensus within the intelligence community was that if the activity was linked to an ABM system, then it was a primitive one. Although an October 1963 NIE judged that the Soviets were deploying ABM around Leningrad, a follow-on estimate in 1964 hedged that conclusion. By 1967, McNamara testified that he believed the system was designed for air defense.[79]

As analysts traded interpretations of the Tallinn Line, Soviet ABM research continued. In January 1964, a 150-foot-tall antenna operated by the Naval Research Laboratory at Chesapeake Beach, Maryland, intercepted the first signals of a new type of Soviet phased array radar, soon dubbed Hen House, near the Sary Shagan test range. Analysts concluded that the radar, still in development, was designed to support an ABM system. A CIA facility in Palo Alto, California, began monitoring Soviet radar tests the following August.[80]

Operational deployment of the Hen House radars began in 1964. Reconnaissance photographs and military attaché reports indicated that the Soviets were constructing two of the radars to the northwest of Moscow.

They were building another radar, code named Dog House, southwest of Moscow and oriented toward the approach path of U.S. ICBMs. Photographs of the radars, together with their electronic signatures, led analysts to believe that they were part of an ABM system.[81] At a news conference in November 1966, McNamara announced that there was "considerable evidence" that the Russians were deploying an ABM system.[82]

In fact, the Soviet leadership had authorized the deployment of an ABM system to protect Moscow in April 1958. The system was designed to intercept missiles outside the earth's atmosphere. Designers envisioned a system comprised of a command center, eight early warning radars, and thirty-two (later reduced to sixteen) battle stations comprising tracking and guidance radars and eight interceptor launchers. Technical difficulties prevented designers from meeting the initial deployment schedule, which envisioned the system becoming operational in 1967. In fact, it was not until 1972 that the first section of the system became operational, with the second operational in 1974.[83]

The system's Hen House radars would provide the initial detection and tracking of inbound ballistic missiles, while the Dog House radar would provide accurate trajectory information. The Try Add radars associated with each battle station would provide accurate tracking of targets and track and guide nuclear-armed Galosh interceptors to their targets.

A second reason for considering the deployment of an ABM system was progress in the development of U.S. antimissile technology. ARPA sponsored BMD research, including Project Defender, which included research into esoteric technologies such as lasers and particle beams for missile defense as well as such less exotic technologies as phased array radars.[84] It also studied space-based missile defense interceptors as part of Project Bambi (Ballistic Missile Boost Intercept).

The first plan to deploy a U.S. ABM system utilized the Western Electric/McDonnell-Douglas LIM-49 Nike Zeus. In development since 1955, the missile was a relatively straightforward development of the MIM-14 Nike Hercules SAM and employed the same command guidance and nuclear warhead as the earlier weapon. In January 1958, Secretary of Defense McElroy decided to give the Army the lead in ballistic missile defense, but also directed that the Air Force continue the development of ABM radar and command and control systems.[85]

In the fall of 1959, the Army proposed an ABM system composed of thirty-five local defense sectors, nine acquisition radars, and a hundred and twenty missile batteries. The system was to achieve initial operational capability by 1964 and be fully operational by 1969. However, the system had a number of significant weaknesses. Because the Nike Zeus was slow, it had to be fired while an incoming warhead was still outside the atmosphere. As a result, it was vulnerable to even rudimentary countermeasures. Moreover, the system utilized a mechanically steered radar that was incapable of processing information on large number of targets. It could therefore be easily overwhelmed. The Air Force was particularly critical of the system, preferring to rely upon the threat of a nuclear response for deterrence.[86]

The Nike Zeus A was soon followed by the Nike Zeus B, a completely new missile sharing only the guidance method and first stage booster of the Nike Zeus A. The missile was a three-stage solid propellant nuclear interceptor designed to intercept Soviet warheads in outer space before they reentered the atmosphere.[87]

On July 19, 1962, a Nike Zeus missile fired from Kwajalein Atoll intercepted an Atlas D ICBM launched from Vandenberg AFB, California. The interceptor's dummy nuclear warhead passed within two kilometers of the Atlas reentry vehicle (RV). In the next test, on December 22, the interceptor passed within two hundred meters. In the entire series of thirteen tests, which stretched until November 1963, the Nike Zeus achieved nine complete and three partial successes.[88]

Despite these results, McNamara decided against deploying the system. Instead, OSD restructured the program to form the Nike X system, which included the long-range Spartan, the short-range Sprint, and new electronically scanning radars. Spartan interceptors would attack incoming missiles in the upper atmosphere, with Sprint engaging the remainder at an altitude of twenty to thirty miles after the atmosphere had stripped away any decoys. Whereas the radar of Nike Zeus could only track one missile at the time, Nike X featured a phased-array radar capable of tracking multiple targets.[89]

Spartan, a development of the Nike Zeus B, entered flight testing in 1968 at Kwajalein Atoll. Spartan was a longer-range, heavier, and higher-performance interceptor. It was armed with a five-megaton nuclear warhead that was designed to destroy a Soviet RV, not with blast (which was impossible at high altitude) but with X-rays.

Sprint came about as the result of the need for a quick-reaction, last-ditch missile to protect targets against warheads leaking through the Spartan defense as well as SLBMs. In March 1964, Martin Marietta was awarded a contract to develop the cone-shaped missile that accelerated at 100 G, achieving Mach 10 in five seconds. The missile's first launch series included twelve successes, two partial successes, and two failures.

Despite such developments, McNamara remained skeptical about the feasibility of ABM. He continued to believe that the deployment of defensive systems by the United States would trigger a reaction by the Soviets. As two of his aides later wrote, "Any attempt on our part to reduce damage to our society would put pressure on the Soviets to strive for an offsetting improvement in *their* assured destruction forces, and vice versa... This 'action-reaction' phenomenon is central to all strategic force planning issues as well as to any theory of an arms race."[90] McNamara's systems analysts calculated that the Soviets could take any number of measures to offset U.S. attempts to limit damage, including equipping their missiles with MIRVs, deploying penetration aids, and increasing the size of their ballistic missile force.[91]

In fact, the Soviets had begun developing countermeasure systems for their ICBMs in the 1960s. The List (Leaf) system, designed for the SS-9, included inflatable metallic balloons to trick exoatmospheric interceptors as well as subscale decoys for use within the atmosphere. The Soviets also deployed countermeasures as part of their upgrade of the SS-11 to the SS-11 Mod 2 configuration.[92]

Another Soviet attempt to counter a U.S. ABM capability was the Fractional Orbiting Bombardment System (FOBS), which the Soviets referred to as a "global missile." Unlike an ICBM, a FOBS would actually place its warheads into orbit. As a result, FOBS warheads could approach the United States from the south, where there was little to no radar coverage. The Soviets conducted twenty-four tests of the system, a derivation of the SS-9, between December 1965 and August 1971 and accepted it into service in November 1968. Notably, the U.S. intelligence community did not credit the Soviets with a FOBS capability until 1971.[93]

In June 1967, Alain Enthoven sent McNamara a proposal for a draft presidential memorandum supporting of the use of the Nike X system to defend Minuteman fields. However, the Secretary of Defense balked at the uncertain effectiveness of the system. Instead, on September 18, 1967, he announced the Sentinel ABM system. Sentinel changed the focus of ABM from general defense to protection against China's tiny missile arsenal.[94] The system would consist of a number of perimeter acquisition radars located across the northern United States; tracking and engagement radars at thirteen locations in the continental United States, Alaska, and Hawaii; and Spartan and Sprint interceptors.

In November 1967 the Army began conducting preliminary surveys for Sentinel around thirteen cities. However, the system faced growing protest.[95] In part, this represented community opposition to having nuclear-armed interceptors deployed nearby. It was also a manifestation of growing antimilitary sentiment brought on by the Vietnam War. In February 1969, shortly after assuming office, President Richard M. Nixon suspended work on Sentinel. Barely a month later, on March 14, 1969, he replaced Sentinel with a new ABM system, dubbed Safeguard. This system was designed to protect the U.S. ICBM force against ballistic missile and FOBS and comprised twelve Spartan/Sprint installations. The first phase was to include sites at Malmstrom AFB, Montana, and Grand Forks, North Dakota, to protect Minuteman silos; the second phase envisioned sites at Whiteman AFB, Missouri, and F. E. Warren AFB, Wyoming.

The logic of the new ABM architecture was open to question. As Lawrence Freedman has written, "the case upon which the Administration based its arguments for Safeguard... involved shaky estimates based on facile extrapolations, hasty judgments on limited evidence, and stretched extrapolations about those crucial Soviet capabilities about which little was known."[96]

The system also met stiff political opposition. Congress approved funding only after Vice President Spiro Agnew cast the deciding vote on August 6, 1969. In November 1971, Congress approved funding for a site to protect Washington, D.C.

The future of ABM became inextricably linked with efforts to limit the size of U.S. and Soviet offensive nuclear arsenals. In 1972, the United States and Soviet Union signed the Strategic Arms Limitation Treaty, which froze the number of ICBM launchers at existing levels and provided for the addition of new SLBM launchers only after the same number of old ICBM and SLBM launchers were dismantled. The two superpowers also signed the ABM Treaty, which limited each to two ABM sites. On July 3, 1974, the United States and the Soviet Union signed an amendment to the ABM Treaty that reduced the number of permissible ABM sites to one.

In May 1972, the Defense Department suspended construction of the Montana ABM site, which was running nineteen months behind that at Grand Forks. The Grand Forks site reached initial operational capability in April 1975 with twenty-eight Sprint and eight Spartan interceptors; it became fully operational on October 1, with seventy Sprint and thirty Spartan interceptors. The following day, Congress voted to shut it down. The Soviet Union's deployment of MIRVed ballistic missiles convinced some that defense against missile attack was futile. Moreover, the system's radars represented lucrative targets. In February 1976, the Army began shutting the system down after barely five months in operation. The missile site radar (MSR) was turned off, the interceptors disarmed and removed. The acquisition radar continued to operate as an early warning system.

The period from 1961 to 1975 was marked by the ascendancy of the Office of the Secretary of Defense, the civilian defense analysts that largely populated it, and their preferred methodology of systems analysis. The armed services had to compete for resources within relatively fixed budget shares, with the Defense Department's civilian leadership acting as the ultimate arbiter of which programs survived. Although Robert S. McNamara, the architect of the approach, left office in 1968, it remained in place throughout the Cold War.

The dominance of defense civilians introduced a new variable into the development of new technology. In some cases, it led to wise decisions. For example, the development of the F-4 Phantom II as a joint Navy–Air Force program gave the Air Force an outstanding aircraft that it never would have developed on its own.

The approach was not, however, an unalloyed success. The TFX program, for example, offers a vivid example of the limits of such civilian-directed joint weapon systems. Although it yielded the F-111, which eventually served the Air Force well, it did not produce the carrier-based aircraft that the Navy sought.

The weapons designed in the 1950s and developed in the 1960s were created to meet the requirements of conventional and nuclear war with the Soviet Union in Central Europe. Many would be employed in a far different setting, in an irregular war in Southeast Asia.

Notes

1. Allan R. Millett and Peter Maslowski, *For the Common Defense: A Military History of the United States of America*, rev. ed. (New York: Free Press, 1994), 554.
2. George W. Baer, *One Hundred Years of Sea Power: The U.S. Navy, 1890–1990* (Stanford, CA: Stanford University Press, 1994), 372.
3. Alain C. Enthoven and K. Wayne Smith, *How Much Is Enough? Shaping the Defense Program, 1916–1969* (New York: Harper & Row, 1971), 2, 7.
4. Ibid., 6.
5. Thomas D. White, "Strategy and the Defense Intellectuals," *Saturday Evening Post*, May 4, 1963.
6. Curtis E. LeMay, *America Is in Danger* (New York: Funk & Wagnalls, 1968), viii, x.
7. House Subcommittee of the Committee on Appropriations, Hearings on Department of Defense Appropriations for 1969, 90th Congress, 2nd Session (Washington, DC: U.S. Government Printing Office, 1968), 54–55.
8. Enthoven and Smith, *How Much Is Enough?* 136, 138, 148.
9. Millett and Maslowski, *For the Common Defense*, 558.
10. Clarence E. Smith, "CIA's Analysis of Soviet Science and Technology," in *Watching the Bear: Essays on CIA's Analysis of the Soviet Union* (Washington, DC: Center for the Study of Intelligence, 2003), chapter 4.
11. When Soviet designers flew aircraft or missiles, they placed sensors on critical components and radioed their status to the ground so that analysts could measure the weapon's performance. U.S. intelligence agencies had to intercept, de-

code, and analyze them. See David S. Brandwein, "Telemetry Analysis," *Studies in Intelligence* 8, no. 4 (Fall 1964): 21–29.

12. Jeffrey T. Richelson, *The U.S. Intelligence Community*, 2nd ed. (New York: Ballinger, 1989), 256, 259.

13. Wyman H. Packard, *A Century of U.S. Naval Intelligence* (Washington, DC: Office of Naval Intelligence and Naval Historical Center, 1996), 201.

14. NIE–11–67, June 1967, "Soviet Military Research and Development," in *CIA's Analysis of the Soviet Union, 1947–1991*, ed. Gerald K. Haines and Robert E. Leggett (Washington, DC: Center for the Study of Intelligence, 2001), 2.

15. *Sixty Years of Reorganizing for Combat: A Historical Trend Analysis* (Ft. Leavenworth, KS: Combat Studies Institute, U.S. Army Command and General Staff College, January 2000), 26.

16. Ibid., 27.

17. Robert A. Doughty, *The Evolution of U.S. Army Tactical Doctrine, 1946–76* (Ft. Leavenworth, KS: Combat Studies Institute, U.S. Army Command and General Staff College, 1981), 23.

18. Elizabeth J. DeLong, James C. Barnhart, and Mary T. Cagle, *History of the Shillelagh Missile System, 1958–1982* (Redstone Arsenal, AL: U.S. Army Missile Command, 1984), 41, 56.

19. Robert J. Sunell, "The Abrams Tank System," in *Camp Colt to Desert Storm: The History of U.S. Armored Forces*, ed. George F. Hofmann and Donn A. Starry (Lexington: University Press of Kentucky, 1999), 434.

20. DeLong, Barnhart, and Cagle, *History of the Shillelagh*, 96.

21. Kenneth P. Werrell, *The Evolution of the Cruise Missile* (Maxwell AFB, AL: Air University Press, 1996), 24.

22. Wilbur D. Jones Jr., *Arming the Eagle: A History of U.S. Weapons Acquisition Since 1775* (Ft. Belvoir, VA: Defense Systems Management Press, 1999), 353.

23. Ibid., 353.

24. Werrell, *Evolution of the Cruise Missile*, 25, 28.

25. Ibid., 28–29.

26. David R. Mets, *The Quest for a Surgical Strike* (Eglin Air Force Base, FL: Air Force Systems Command Monograph, Armament Division, 1987), 106.

27. Pavel Podvig, ed., *Russian Strategic Nuclear Forces* (Cambridge, MA: MIT Press, 2001), 286–90.

28. Steven J. Zaloga, *The Kremlin's Nuclear Sword: The Rise and Fall of Russia's Strategic Nuclear Forces, 1945–2000* (Washington, DC: Smithsonian Books, 2002), 115.

29. Podvig, *Russian Strategic Nuclear Forces*, 290–94.

30. Cote, *The Third Battle*, 23.

31. Gary E. Weir, "From Surveillance to Global Warming: John Steinberg and Ocean Acoustics," *International Journal of Naval History* 2, no. 1 (April 2003): 4.

32. Cote, *The Third Battle* 25; Weir, "From Surveillance to Global Warming," 3.

33. Ibid., 25.

34. Christopher Ford and David Rosenberg, *The Admirals' Advantage: U.S. Navy Operational Intelligence in World War II and the Cold War* (Annapolis, MD: U.S. Naval Institute Press, 2005), 36–38; Packard, *Century of U.S. Naval Intelligence*, chapter 14.

35. Cote, *The Third Battle*, 39.

36. Ibid., 48.

37. Ibid., 50.

38. Podvig, *Russian Strategic Nuclear Forces*, 294–98.

39. Baer, *One Hundred Years of Sea Power*, 396.

40. Cote, *The Third Battle*, 73.

41. Podvig, *Russian Strategic Nuclear Forces*, 298–302.

42. Cote, *The Third Battle*, 69.

43. "Letter from Secretary of Defense McNamara to President Kennedy," Washington, February 20, 1961, in *Foreign Relations of the United States, 1961–1963*, ed. David S. Patterson (Washington, DC: U.S. Government Printing Office, 1996), 8:35–48.

44. "Memorandum from the President's Military Representative (Taylor) to President Kennedy," Washington, September 19, 1961, in ibid., 126–29.

45. Desmond Ball, "The Development of the SIOP, 1960–1983," in *Strategic Nuclear Targeting*, ed. Desmond Ball and Jeffrey Richelson (Ithaca, NY: Cornell University Press, 1986), 62.

46. "Memorandum from the Chairman of the Joint Chiefs of Staff (Lemnitzer) to Secretary of Defense McNamara, CM–190–61," Washington, April 18, 1961, in Patterson, *Foreign Relations of the United States*, 74–78.

47. Ball, "Development of the SIOP," 63.

48. Ibid., 65.

49. "Draft Memorandum from Secretary of Defense McNamara to President Johnson," Washington, December 6, 1963, in Patterson, *Foreign Relations of the United States*, 549.

50. Ball, "Development of the SIOP," 69.

51. Ibid.

52. "Policy for Planning the Employment of Nuclear Weapons," National Security Council, NSDM-242, January 17, 1974, Washington, DC.

53. Ball, "Development of the SIOP," 74.

54. A. W. Marshall, *Long-Term Competition with the Soviets: A Framework for Strategic Analysis* (Santa Monica, CA: Rand Corporation, 1972), vi.

55. Ibid., vii.

56. Ibid., viii.

57. Zaloga, *The Kremlin's Nuclear Sword*, 77, 68, 70.

58. Frank Eliot, "Moon Bounce Elint," *Studies in Intelligence* 11, no. 2 (Spring 1967): 59–65; Albert D. Wheelon, "Technology and Intelligence," *The Intelligencer: Journal of U.S. Intelligence Studies* 14, no. 2 (Winter/Spring 2005), 51–56.

59. Packard, *Century of U.S. Naval Intelligence*, 52.

60. NIE 11–8-64, October 1964, *Soviet Capabilities for Strategic Attack*, in Haines and Leggett, *CIA's Analysis of the Soviet Union*, 2.

61. Lawrence Freedman, *U.S. Intelligence and the Soviet Strategic Threat*, 2nd ed. (Princeton, NJ: Princeton University Press, 1986), 153.

62. Ibid., 114.

63. Robert L. Hewitt, John Ashton, and John H. Milligan, "The Track Record in Strategic Estimating: An Evaluation of the Strategic National Intelligence Estimates, 1966–1975," unpublished paper, February 6, 1976, iii.

64. Freedman, *U.S. Intelligence*, 104.

65. Soviet sources disagree over the purpose of the SS-9. Some argue that the missile was designed specifically for the destruction of U.S. ICBM fields (see

Podvig, *Russian Strategic Nuclear Forces*, 196); others argue that the SS-9 was not a counterforce weapon (see Zaloga, *The Kremlin's Nuclear Sword*, 132).

66. Freedman, *U.S. Intelligence*, 109.
67. Smith, "CIA's Analysis."
68. Freedman, *U.S. Intelligence*, 135.
69. Smith, "CIA's Analysis."
70. Peter Roman, "Strategic Bombers Over the Missile Horizon, 1957–1963," *Journal of Strategic Studies* 8, no. 1 (March 1995): 198–208.
71. Freedman, *U.S. Intelligence*, 98.
72. Jacob Neufeld, *Ballistic Missiles in the United States Air Force, 1945–1960* (Washington, DC: Office of Air Force History, 1990), 230.
73. Freedman, *U.S. Intelligence*, 98.
74. Ibid., 99.
75. Donald MacKenzie, *Inventing Accuracy: A Historical Sociology of Nuclear Missile Guidance* (Cambridge, MA: MIT Press, 1990), 213.
76. Quoted in Lawrence Freedman, *The Evolution of Nuclear Strategy* (New York: St. Martin's, 1981), 252–53.
77. Freedman, *U.S. Intelligence*, 83.
78. Donald R. Baucom, *The Origins of SDI: 1944–1983* (Lawrence: University Press of Kansas, 1992), 30.
79. Freedman, *U.S. Intelligence*, 93.
80. Eliot, "Moon Bounce Elint."
81. Wheelon, "Technology and Intelligence," 56.
82. Freedman, *U.S. Intelligence*, 88.
83. Podvig, *Russian Strategic Nuclear Forces*, 413–14, 416.
84. Baucom, *The Origins of SDI*, 16.
85. Ibid., 11.
86. Ibid., 9.
87. Ibid., 7.
88. Ibid., 17.
89. Ibid., 19.
90. Enthoven and Smith, *How Much Is Enough?* 176.
91. Ibid., 188.
92. Zaloga, *The Kremlin's Nuclear Sword*, 128–29.
93. Ibid., 113.
94. Freedman, *U.S. Intelligence*, 125.
95. Ibid., 129.
96. Ibid., 146–47.

3

Technology and the War in Vietnam, 1963–1975

THE VIETNAM WAR demonstrated the limits of American power in general, as well as the limitations of the American military's reliance on technology in particular. A standard critique is that the United States fought the war as if it were a high-intensity conflict against the Soviet Union across Germany's Fulda Gap. The Army, for example, emphasized conventional operations over counterinsurgency, whereas the Air Force employed aircraft and tactics designed for Central Europe to wage irregular warfare in the jungles of Southeast Asia.[1]

As is often the case, the conventional wisdom is only half-right. The Vietnam War was a peripheral campaign that occurred within the context of the Cold War, and the U.S. military fought it that way. It was the prospect of a war with the Soviet Union, not the conflict in Southeast Asia, that drove the U.S. armed services' requirements. At the same time, the services adopted innovative technology in a number of areas. The Army deployed airmobile divisions that, though originally designed for the nuclear battlefield, gave U.S. troops an edge in firepower and mobility over their adversaries. The Navy, for its part, fielded a riverine force to interdict communist supply lines, while the Air Force modified transport aircraft into gunships to perform the same mission. The war also saw the introduction of a range of new technologies, including unmanned aerial vehicles (UAVs), unattended ground sensors, and precision-guided munitions (PGMs), that would prove their worth in the wars of the 1990s and beyond.

Technology and the Vietnam War

The Vietnam War was a classic example of the American way of war. The commander of U.S. forces in Vietnam, General William Westmoreland, pursued a strategy that Grant, Pershing, or Eisenhower would have been comfortable with, one based on the massive application of technology and firepower to launch offensive operations to annihilate the enemy. As Russell Weigley put it, "the great world wars and the military history that had preceded them had so conditioned American military thought that their influence could not be escaped however different the circumstances of new combats might be."[2]

Air Combat

Because air combat is one of the most technologically intensive realms of warfare, it offers the best case for examining the role that technology played in U.S. operations in Vietnam. It was also in the air that the two sides were most evenly matched: throughout the war, state-of-the-art U.S. aircraft faced a frontline Soviet air defense network. Although technology contributed to the effectiveness of U.S. air forces, it was not the only or even the most important ingredient to its success. It was only when technology was combined with innovative tactics and organizations that it had a real impact. In other cases, new technologies did not realize their potential because they were immature or fell prey to organizational resistance.

The Air Force was in many ways unprepared for a limited conventional war in Southeast Asia. Many Air Force leaders felt that forces designed for a nuclear war against the Soviets would be adequate for war in Vietnam. During Operation Rolling Thunder between 1965 and 1968, the majority of the service's strike missions against heavily defended targets in North Vietnam were flown by the Republic F-105 Thunderchief, an aircraft originally designed for air combat but later modified as a nuclear bomber for a war against the Warsaw Pact in Central Europe. The fastest aircraft in the world at low altitude, it was also unmaneuverable: early in its career it had acquired a string of uncomplimentary nicknames, such as the "Lead Sled," "Ultrahog," and "Thud." Nor did it fare well in air-to-air combat, sustaining the highest loss rate of any U.S. aircraft in Southeast Asia. All told, the Air Force lost 332 of them.[3]

At the time of the first U.S. air strikes, North Vietnam had a fairly simple air-defense network composed of fewer than a hundred aircraft, none of which were jet-powered; few early warning radars; and no SAMs. However, that situation quickly changed: two days after U.S. air strikes began, U.S. reconnaissance photos showed MiG-17 Fresco jet fighters at an airfield near Hanoi.[4]

On paper, U.S. Air Force F-105s and Navy F-4 Phantom IIs and F-8 Crusaders and U.S. Navy A-4 Skyhawks and A-6 Intruders were far superior to the MiG-17, an upgraded version of the Korean War-vintage MiG-15 Fagot. The U.S. fighters were supersonic and equipped with air-to-air missiles (AAMs), whereas the MiG-17 was subsonic and had cannons. However, the MiG-17 could outturn and outperform U.S. jets at subsonic speed in horizontal flight and thus possessed a significant dogfighting advantage.[5]

Countering North Vietnamese SAMs

By mid-1965, U.S. intelligence estimated that North Vietnam had more than doubled its inventory of early warning radars. In April, it gathered the first evidence of the presence of SA-2 Guideline SAMs in North Vietnam. The large radar-guided missile was designed to shoot down high-flying bombers like the B-52. A typical site consisted of several long-range early warning radars, a FAN SONG guidance radar, and six missile launchers.[6] Between July 1965 and March 1968, the North Vietnamese fired between 5,366 and 6,037 SAMs at U.S. aircraft.[7]

U.S. forces developed both technological and tactical responses to North Vietnamese SAMs. First, the Air Force deployed EB-66 electronic warfare aircraft to support strike missions. The two-engine EB-66, developed from the B-66 Destroyer bomber, performed two missions: the EB-66B electronic reconnaissance aircraft identified North Vietnamese radars, while the EB-66B and EB-66E aircraft used electronic countermeasures to jam them. The two types of aircraft often operated in tandem.[8]

The Air Force and Navy also began equipping their aircraft with ECM. In July 1965, the Air Force began outfitting its RF-101 Voodoo and RF-4 reconnaissance aircraft with QRC-160 jamming pods, based upon countermeasures first developed for strategic bombers. Each pod contained four jammers: two for use against the FAN SONG radar and two against radars used to guide anti-aircraft artillery (AAA). In September 1966, the first twenty-five pods arrived in theater; by the beginning of 1967, there were fifty-one in Southeast Asia. The Navy, for its part, mounted ALQ-51 ECM packages internally in its A-4 and A-6 attack aircraft.[9]

By the beginning of 1967, the deployment of these ECM systems had reduced aircraft losses significantly. As one air-wing commander put it, "Seldom has a technological advance of this nature so degraded the enemy's defensive posture. It has literally transformed the hostile defense environment we once faced, to one where we can now operate with a latitude of permissibility."[10]

The Air Force also deployed radar homing and warning (RHAW) equipment on its F-105s to alert pilots to a SAM launch. Taking advantage of the fact that it took the Fan Song time to acquire a target, lock on, and fire a SAM, RHAW equipment gave pilots time to take evasive action, deploy chaff, and activate their jammers.[11]

U.S. air forces also adapted tactically to the SAM threat. Strike formations began conducting low-level attacks to stay below the engagement envelope of the SA-2. However, these measures created new vulnerabilities: while low-altitude attack limited losses to SAMs, it dramatically increased losses to AAA.[12]

Finally, U.S. air forces developed tactics to attack SAM sites using so-called Iron Hand missions. The Navy initially used A-4 Skyhawk aircraft for the mission, while the Air Force used two-seat F-100F Super Sabres and later F-105F and F-4Cs.[13] These "Wild Weasel" aircraft would precede a strike formation into the target area, attempting to knock out or neutralize North Vietnamese SAMs before the main attack arrived.

The Navy began to deploy guided weapons to suppress North Vietnamese SAMs. In March 1966, the Navy introduced the AGM-45A Shrike antiradiation missile (ARM), which was designed to home in on the Fan Song's emissions. Although the missile enjoyed some success, it had limited range and maneuverability and could be spoofed. As the North Vietnamese adapted, its kill rate fell from 28 percent in 1966 to 18 percent in 1967. In March 1968, the Navy introduced the AGM-78 Standard ARM. It had longer range and a heavier warhead than the Shrike, and had a computerized memory that allowed it to home in on a radar even after it shut down. On the other hand, it cost ten times as much as the earlier missile and proved unreliable, with some 30 percent failing to launch.[14]

Although the introduction of ARMs helped reduce the SAM threat, they were not the solution. Indeed, it was only the adoption of both technology (such as ECM and ARMs) and tactics (such as Iron Hand missions) that helped U.S. aviators minimize the SAM threat. In 1965, North Vietnamese SAMs shot down one aircraft for every sixteen missiles fired. In 1966, that ratio declined to one aircraft for every thirty-three missiles. By 1967, it was down to one aircraft per fifty missiles, and by 1968 it took more than a hundred missiles to down a single aircraft. Although the North Vietnamese responded by changing their SAM tactics, U.S. aircraft were still able to resume operating at medium altitude. This protected them from AAA, improved their ability to navigate to targets, and permitted better bombing accuracy.[15]

Countering North Vietnamese Fighters

In 1966, the appearance in the skies over North Vietnam of the Soviet Union's most modern fighter, the MiG-21 Fishbed, complicated the battle for air superiority over North Vietnam. The aircraft was more maneuverable and had greater acceleration than the U.S. F-4 and F-8 and was armed with K-13 (AA-2 Atoll) air-to-air missiles. North Vietnam also began deploying an increasingly sophisticated ground-control intercept (GCI) system to direct its fighters.[16]

The greatest challenge U.S. fighters faced was the need to spot North Vietnamese interceptors before they ambushed U.S. strike aircraft. Because U.S. aircraft operating over North Vietnam initially lacked GCI radar support, North Vietnamese fighters often surprised them. To help remedy this deficiency, in July 1966 the Navy established Red Crown patrols by ships equipped with powerful air-search radars off the mouth of the Red River.[17] In addition, the Air Force deployed EC-121D Big Eye airborne radar surveillance aircraft to Thailand. These radars allowed controllers to spot North Vietnamese aircraft and warn U.S. aircrews. To assist the EC-121 crews in distinguishing U.S. from North Vietnamese aircraft, U.S. aircraft were equipped with transponders that allowed air controllers to identify them as friendly. However, such innovations met cultural resistance: U.S. crews frequently turned these off because they believed (incorrectly) that the North Vietnamese could also interrogate them. In other cases, pilots shut them off to prevent Big Eye from tracking them while they were operating in restricted areas.

Although the North Vietnamese could not interrogate U.S. identify friend or foe (IFF) transponders, the United States developed just such a capability against North Vietnamese transponders. In early 1967 the United States deployed the QRC-248 IFF transponder interrogator, which was able to read the signals transmitted by the SRO-2 transponder on Soviet export fighters. The development was a windfall, since the North Vietnamese GCI net depended heavily upon the transponders. By the end of May 1967, all the Air Force's EC-121D College Eye aircraft had received the equipment.[18]

Information from the QRC-248 proved to be accurate and reliable, providing U.S. forces warning of North Vietnamese interceptors. The device was most effective when it actively interrogated enemy IFF transponders. However, the National Security Agency and Joint Chiefs of Staff were concerned that using the device in that mode could compromise its existence, so they restricted its use to eavesdropping on North Vietnamese transponders until late July 1967.[19]

In January 1967, the Air Force began deploying an IFF interrogator, the APX-80 Combat Tree, aboard some F-4Ds. Like the QRC-248, it was designed to trigger North Vietnamese IFF transponders. Although only five Combat Tree–equipped aircraft were deployed to the theater by May 1972, the aircraft had a major effect: over the next five months, they were responsible for seventeen of the Air Force's twenty aircraft kills[20]

In the summer of 1967, the Air Force introduced the EC-121K Rivet Top. Originally an electronic intelligence aircraft, it was modified to identify SAM sites and warn of attack. It also contained four stations where South Vietnamese linguists monitored the transmissions of North Vietnamese air defense networks. It carried the QRC-248, but also equipment to interrogate two other types of transponders. It was so successful that it led to another program, Rivet Gym, that added linguist stations to EC-121Ds.[21]

In July 1972, the U.S. established a fusion center that monitored North Vietnamese MiG activity, using the code name Teaball, at Nakhon Phanom, Thailand. The center fused a variety of sources, including sensitive information derived from communication intelligence. While it gathered considerable information of use to U.S. aircrews, the problem of how to use the information they gathered while also protecting its source proved to be a vexing one. Because Teaball could not disclose the source of its information, aircrews were initially dubious of the information. It was also hampered by the fact that it could not communicate directly with strike aircraft.[22] Over time, however, pilots began to rely on Teaball data.

Air-to-Air Combat

At the beginning of the air war, many predicted that AAMs would give the United States a major edge. AAMs such as the AIM-7 Sparrow and AIM-9 Sidewinder were thought to be so effective that the F-4 Phantom II was designed without a cannon. In fact, U.S. AAMs proved to be one of the most significant technological disappointments of the war.

There are several reasons why U.S. AAMs faired so poorly in combat. First, they were designed against high-altitude, nonmaneuvering targets such as Soviet intercontinental bombers, not aircraft maneuvering at low altitude. Second, they were designed for operations from the continental United States or European bases, not for conditions in Southeast Asia.

Third, manhandling in theater and repeated rough takeoffs and landings often damaged the missiles' sensitive electronics.

The medium-range radar-guided AIM-7 Sparrow was the most frequently used AAM in the Vietnam War. Prior to the war, the Defense Department expected it to have a 70 percent probability of kill. In fact, during Operation Rolling Thunder, only 8 percent of Sparrow launches resulted in hits, leading to fifty-six kills. The missile often simply failed to function properly due to its complexity and a large amount of sensitive equipment.[23] In addition, pilots were poorly trained and often launched missiles out of parameters. Out of some six hundred missiles fired between 1965 and 1973, only 10 percent had any chance of hitting their targets.[24] The need for positive visual identification of targets severely restricted the possibility of long-range engagements. Indeed, U.S. airmen apparently achieved only two beyond-visual-range kills in the entire war.[25]

The AIM-9 Sidewinder was the cheapest, simplest, and most effective AAM employed by the United States in the Vietnam War. Although about twice as lethal as other AAMs, it was far from an absolute success: only 15 percent of *Sidewinder* launches were kills, leading to the destruction of eighty-one aircraft.[26] By contrast, prewar testing predicted that the missile would hit 60 percent of the time.[27] The missile could easily be spoofed and was easy to fire outside its engagement envelope.

The AIM-9E, the Air Force's upgrade of the AIM-9B, was if anything a step backward in terms of effectiveness: the missile achieved a 12 percent probability of kill, compared to 15 percent for the AIM-9B. The AIM-9G, the Navy's upgrade of the AIM-9B, was so clearly superior that the Air Force requested the missile. However, it was incompatible with Air Force launch rails and electronics. In January 1972, the Air Force restarted the AIM-9J program. However, the missile did not appear in combat until the end of July 1972.[28]

In the first months of 1967, the Air Force and Navy achieved lopsided results against the North Vietnamese air force. Between January and March, American fighters shot down twelve North Vietnamese fighters without sustaining losses. In April and May, U.S. forces downed thirty-eight MiGs while losing eight aircraft. The North Vietnamese then stood down and regrouped, taking to the air again in August with substantially revised tactics and increased emphasis on ground-controlled intercepts using the MiG-21. Thereafter, U.S. fortunes worsened: in the first three months of 1968, 22 percent of U.S. fighters were lost in air-to-air combat.[29]

The Air Force and Navy took divergent approaches to improving their effectiveness in the air. The Air Force focused on technical solutions. It deployed the F-4E, which was equipped with an internal M-61 cannon and an improved radar system that made the gun very accurate. In late 1972, it began deploying modified F-4Es, code-named Rivet Haste, with leading edge slats to improve the aircraft's maneuverability, Combat Tree, and a long-range telescope, called Target Identification System, Electro-Optical (TISEO) to allow pilots to identify visually hostile aircraft at longer range. Rivet Haste crews also received special training in air-to-air combat. [30] However, such crews arrived in theater too late to affect the war.

The Navy's reconsideration of fighter tactics was much more fundamental and focused on tactics more than technology. The Navy focused upon improving the tactical skills of its pilots, primarily through the establishment of the Naval Fighter Weapon School Gun at NAS Miramar in San Diego, CA, also known as "Top Gun."[31] Top Gun's curriculum was designed to hone the dogfighting skills of some of the Navy's best aviators. The Navy used A-4 aircraft to simulate the MiG-17 and T-38 aircraft to simulate the MiG-21. Improved dogfighting skills showed in the skies over Vietnam: from May 1972 on, the Navy's kill ratio climbed to 8.33:1. Moreover, most kills were by Top Gun graduates.

Air Force training, by contrast, remained stagnant. Not surprisingly, the Air Force's kill ratio improved only slightly as the conflict wore on. It was not until 1975 that the Air Force began its own dogfighting exercises under the name Red Flag.

In developing new tactics, U.S. pilots were in some cases able to fly against actual Soviet combat aircraft obtained by the U.S. intelligence community. In 1967, the Defense Intelligence Agency (DIA) acquired a MiG-21 from a foreign source. The Navy and Air Force used the aircraft for a classified flight-test program known as Have Doughnut. In 1969, the United States reached an agreement with Israel to acquire another MiG-21 and two MiG-17s. The latter were tested in a project known as Have Drill.[32] These programs exposed the weaknesses of the Soviet fighters and helped pilots develop tactics to exploit them.

Strategic Bombing

SAC's strategic bombers also saw action in the war, albeit performing missions very different from those their crews had been trained to execute. Throughout the Cold War, SAC crews trained to conduct single-aircraft, low-altitude, high-speed attacks to avoid Soviet SAMs and interceptors on their way to delivering their nuclear payloads. By contrast, conventional bombing required large formations of bombers to operate at high altitude. SAC bomber crews were unaccustomed to formation flying and pattern bombing, and the command lacked conventional munitions and delivery systems.[33]

The use of nuclear bombers in a conventional role also went against the command's culture. The SAC leadership argued against the deployment of B-52s to Southeast Asia, claiming that it would detract from the SIOP, require the reconfiguration of aircraft, and could lead to the compromise of sensitive technologies. As the SAC commander, General Thomas Power, put it, "[don't] talk to me about that; that's not our life. That's not our business. We don't want to get in the business of dropping any conventional bombs. We are in the nuclear business, and we want to stay there."[34] SAC similarly resisted deploying its tankers and electronic warfare aircraft to Southeast Asia.[35]

Despite such objections, the first thirty B-52s deployed to Guam in February 1965. To improve their conventional capability, the Air Force added external bomb racks, giving each bomber the ability to deliver almost twenty-six tons of bombs per sortie. In December 1965, the Air Force began modifying the bomb bay of some B-52s, increasing their capacity to eighty-four 500-pound or forty-two 750-pound bombs, giving these "Big Belly" B-52s the ability to deliver thirty tons of ordnance per sortie.[36]

The Air Force established Tactical Air Navigation (TACAN) stations throughout Southeast Asia to guide strike aircraft. To increase the precision, and hence the effectiveness, of bombing missions, in early 1965 SAC began to study the feasibility of a radar-based strike direction system. The result was the MSQ-77 ground-based radar system. In tests in late 1965 and early 1966, bombers using the MSQ-77 achieved a 486-foot CEP.[37] Under the code name Combat Skyspot, SAC established six sites—five in South Vietnam and one in Thailand—to direct strikes across all South Vietnam, southern Laos, and southern North Vietnam. Under the code name Heavy Green, between October 1967 and March 1968 the Air Force operated a Combat Skyspot station in northeast Laos to permit radar-directed bombing in Laos and North Vietnam.[38] The secret site allowed the U.S. military to bomb targets in and around Hanoi in all weather, until March 11, 1968, when a specially trained North Vietnamese sapper unit destroyed the site in a well-planned attack.

U.S. air strikes had limited effect on the communists. As one CIA assessment put it, "Despite continual aerial reconnaissance and air strikes, the North Vietnamese have supported insurgent wars in four countries and withstood daily bombardment of supplies and facilities within their own country."[39] The North Vietnamese and Vietcong were masters of denial and deception, camouflaging air defense sites, aircraft and air installations, naval combatants and merchant ships, radar and communication facilities, and military bases.

Gunships

Although much of the air fleet employed over North Vietnam was designed for a nuclear war with the Soviet Union, in several cases the U.S. Air Force developed innovative aircraft to prosecute the war. The most prominent case was that of the gunship. It is significant, however, that the idea came not from the Air Force's research and development bureaucracy, but from Ralph Flexman, an Air Force Reserve major serving on active duty. Flexman proposed arming an aircraft with side-firing cannons, which would allow it to concentrate its fire on ground targets while executing a slow turn. In mid-1963, the Air Force undertook a modest test program, which led to a requirement for a C-47 gunship fitted with three side-mounted 7.62 mm Gatling guns. The aircraft, nicknamed "Puff the Magic Dragon" and later given the call sign "Spooky," flew low and slow and provided both on-call firepower and interdiction of communist logistics. Eventually, fifty-three of the transport aircraft were converted to gunships; seventeen were lost in combat and two were lost in non-combat circumstances.[40]

Gunships received the enthusiastic support of the secretary of defense. They were also popular with soldiers and marines. But even though the service's secretary supported them, they were not accepted by parts of the Air Force. Opposition to the gunships came from the Tactical Air Command, which questioned the survivability of low-flying, slow-moving aircraft. Moreover, gunship pilots were hardly mainstream. In a service where bomber (and increasingly fighter) pilots dominated the leadership, gunship pilots were marginal. The gunship was a niche capability for a peripheral war.

In November 1966, the Air Force chose the AC-130 Spectre as the replacement for the AC-47. The aircraft was able to carry much more firepower than that it replaced, including four 7.62 mm and four 20mm Gatling guns. It also had a night vision scope, side- and forward-looking radar, and forward-looking infrared sensors. It possessed a computer fire control system, two steerable 20-kilowatt lights, a semiautomatic flare dispenser, and armor protection. It could spend literally hours orbiting over the battlefield.[41]

The AC-130 underwent a number of upgrades in subsequent years. The first, code named Surprise Package, gave the aircraft a standard armament of two 20 mm and two 40 mm guns, low-light television and improved infrared sensors, a new digital computer, and a laser designator. The first of these modified aircraft arrived in theater in December 1969. Project Pave Aegis modified the AC-130 to mount two 20 mm, one 40 mm, and one 105 mm howitzer.[42]

Technology and Culture

Technology clearly played an important role in the air war over Vietnam. Communication intelligence, electronic countermeasures, and antiradiation missiles all helped the U.S. Navy and Air Force operate over North Vietnam. Such technologies were, however, insufficient to yield tactical, let alone strategic, success. It is difficult, however, to envision a scenario in which air power could have proved decisive in a war that was fundamentally a struggle for the support of the people.

Technology was refracted through the prism of the Air Force's organizational culture throughout the war. Because bomber and increasingly fighter pilots dominated the Air Force, air superiority and strike aircraft received the greatest emphasis. This was reinforced by the fact that the Air Force's central planning contingency was a confrontation with the Soviet Union in Central Europe. Capabilities of greater importance to Vietnam, such as gunships, were consigned to niche roles. Indeed, when an act of Congress created U.S. Special Operations Command in 1986, the Air Force's gunships were transferred to the new command's air component.

Vietnam had a profound influence on the structure of the Air Force. Because the demand for pilots in Vietnam far exceeded the supply, the conflict broke down some of the barriers that had existed between different Air Force communities. A number of SAC pilots, for example, joined the tactical air force, where they encountered a much more decentralized and innovative culture. More significantly, the need for large numbers of fighters to support strike operations in Vietnam gave fighter pilots opportunities to excel in battle and ultimately advantages when it came to promotion. As the war wore on, fighter pilots began to break into the top echelon of the service, displacing the bomber pilots.[43] Indeed, every Air Force chief of staff since General Lew Allen Jr., who served from 1978 to 1982, has been a fighter pilot.

The Army and Airmobility

Although ground combat is less technologically intensive than war in the air or at sea, technology nonetheless played a significant role in the U.S.

Army's conduct of the war in Vietnam. The Army's approach to the war, which Andrew Krepinevich has dubbed the "Army Concept," emphasized the profligate use of firepower to find, fix, and destroy the communists.[44] The attrition strategy played to America's material abundance and techno-logical superiority as well as its aversion to casualties. However, firepower also killed innocents and undermined efforts to win the allegiance of the Vietnamese people.

By and large, the Army fought the war with conventional Reorganiza-tion Objective Army Division (ROAD) units designed for Europe's Cen-tral Front. The service's primary innovation was the introduction of air-mobile formations. As one Army study put it, "The widespread use of the helicopter was the most significant advance of the Vietnam War." In the words of the Army chief of staff (and later chairman of the Joint Chiefs of Staff), General Earl Wheeler, "helicopters and offensive airpower provide friendly forces with advantages in mobility and firepower greatly exceed-ing those available to counterinsurgency forces in any other anti-guerrilla war in history. And… mobility and firepower are the fundamental keys to success in combat."[46]

Although airmobile units saw their combat debut in Southeast Asia, their genesis lay in the need to concentrate forces rapidly on the nuclear battlefield. Writing in *Harper's* magazine in April 1954, Major General James M. Gavin argued that only through the widespread adoption of helicopters could the U.S. Army be effective on the modern battlefield. As a veteran of the campaigns in Sicily, Salerno, Normandy, and Holland, Gavin spoke with considerable authority. In his view, the mechanization of the Army had decreased its ability to perform traditional cavalry func-tions:

> Cavalry is not a horse, nor the crossed sabers and yellow scarves. These are the vestigial trappings of a gallant great arm of the U.S. Army, whose soul has been traded for a body. It is the arm of Jeb Stuart, and Custer, and Sheridan, and Forrest. It is the arm that as late as World War II got there (in Forrest's phrase) the "fustest with the mostest" but is now rapidly becoming, in terms of firepower and mobility, lastest with the leastest…. With the motorization of the land forces, and the consequential removal of the mobility differential, the cavalry has ceased to exist in our Army except in name.[47]

On the modern battlefield, he argued, cavalry functions needed to be performed more rapidly and at a greater distance than had heretofore been possible. Only through the widespread use of helicopters could the Army pursue the traditional cavalry roles of screening, reconnaissance, exploitation, and pursuit. Just as tanks had replaced horses, Gavin now argued that tanks should succeed tanks. Gavin thus made the case for radical change—the development of airmobile units—through an appeal to traditional army missions.

In June 1956, Brigadier General Carl I. Hutton, the commandant of the Army Aviation School, organized a series of tests of armed helicopters. He gave Colonel Jay T. Vanderpool the assignment of developing a "fighting helicopter." In two weeks, Vanderpool's team of five armed a Bell H-13, the smallest helicopter available, with two .50 caliber World War II aircraft guns and launch rails for 80mm rockets. They then assembled an experimental company-size air cavalry organization manned by military and civilian volunteers. The unit tested various types of ordnance and developed air cavalry tactics, culminating in a demonstration in 1957. It also wrote the Army's first air cavalry manual, drawing heavily on a 1936 cavalry manual as a way of portraying the new organization in terms that were intelligible to senior officers.[48]

Army leaders saw mobility as vital to the ability of the United States to fight numerically superior enemies. As Secretary of the Army Wilber Brucker wrote in 1956, "Tactical victory will belong to the army with the superior mobility on a rolling battlefield; in the sense defined by future wars, aviation is a most important form of mobility."[49]

Despite such rhetoric, Secretary of Defense McNamara was frustrated by what he saw as the Army's lack of progress in air mobility. As he put it in an April 19, 1962, memorandum to the Army staff:

> I have not been satisfied with Army program submissions for tactical mobility. I do not believe the Army has fully explored the revolutionary opportunities offered by aeronautical technology for making a revolutionary break with traditional surface mobility means. Air vehicles operating close to, but above the ground appear to me to offer the possibility of a quantum increase in effectiveness.... I therefore believe that the Army's reexamination of its aviation requirements should be a bold "new look" at land warfare mobility. It should be conducted in an atmosphere divorced from traditional viewpoints and past policies.... It also requires that bold, new ideas which the task force may recommend be protected from veto or dilution by conservative staff review.[50]

In response, the Army established the Army Tactical Mobility Requirements Board (also known as the Howze Board after its head, General Hamilton Howze). In less than four months, the board issued a 3,500-page report recommending the establishment of airmobile units in which helicopters and fixed-wing aircraft would replace many surface vehicles.[51] It recommended establishing at least five air assault divisions, which would be similar to infantry divisions but with attack and transport aircraft and helicopters replacing artillery and vehicles. It also called for the Army to field three air cavalry combat brigades, each with 144 antitank helicopters, and five air transport brigades, each with eighty fixed-wing and fifty rotary-wing aircraft.[52]

The Howze Board's recommendations were bold, calling for a force that would rely heavily on aircraft that were either recently deployed or still in development, such as the OH-6A Cayuse and OH-58A Kiowa helicopters and AO-1 Mohawk and AC-1 Caribou aircraft. It also proposed roughly doubling the number of aviators and aircraft in the by 1968.[53]

Incoming Secretary of the Army Cyrus Vance and Army Chief of Staff Wheeler strongly endorsed the report. Secretary of Defense McNamara approved increasing troop strength and the activation of a provisional air assault division and air transport brigade for further test and evaluation.[54] On February 15, 1963, the Army established the 11th Air Assault Division (Test) and the 10th Air Transport Brigade at Ft. Benning, Georgia. Brigadier General Harry Kinnard was placed in command and given orders "to determine how far and how fast the Army can go, and should go, in embracing airmobility."[55]

In the spring of 1965, the Army was tasked with sending a unit to Vietnam's Central Highlands. None of its existing formations seemed suitable: airborne forces had limited mobility, while armored and infantry forces were too heavy and dependent on vehicles. As a result, on July 1, 1965, General Harold Johnson activated the 1st Cavalry Division (Airmobile) with resources from the 11th Air Assault Division (Test) and 2nd Infantry Division and gave it ninety days to begin deploying to Vietnam.[56] By October 3, the entire division was located at its base area at An Khe. By mid-October, the North Vietnamese Army began building up forces and assembling three regiments in the area.

On November 14, some 450 men of the 1st Battalion, 7th Cavalry Regiment landed by helicopter in a small clearing in the Ia Drang Valley. Two thousand North Vietnamese soldiers surrounded them. The result was the first major battle of the Vietnam War. Over the course of three days, the soldiers withstood repeated communist assaults and eventually took the battlefield. Throughout the battle, helicopters provided fire support, resupplied the troops, and evacuated the wounded.

Throughout the war, helicopters flew more than 36 million sorties, including 7.5 million assault sorties, and nearly 5,000 were lost to enemy fire or accidents.[57] The war saw helicopters used for a variety of roles. For example, some Hueys were equipped with banks of radios so that they could act as heliborne command posts, allowing commanders to control forces from high above the battlefield.[58] The director of the Joint Research and Testing Agency in Vietnam viewed the Heliborne Command Post as "the single piece of new materiel which should have the most influence on improving the conduct of the war in Vietnam."[59] In 1967, the Army introduced the UH-1B Huey gunship to provide fire support to airmobile forces. Armed with machine guns, 7.62mm miniguns and 2.75-inch rockets, it was the Army's primary attack helicopter throughout the war.

Helicopters gave U.S. forces a marked advantage in mobility. They allowed U.S. forces to fix communist forces in place; when they did, U.S. forces were able to defeat them. Often, however, communist forces were able to melt away when confronted. The communists employed lookouts on trees to warn of approaching helicopters. If they did not want to fight, they simply avoided U.S. forces and reoccupied the area after the Americans left.

The development of heliborne units proved to be the most enduring of the Vietnam-era organizational innovations. Airmobile and air assault units remain part of the Army's force structure three decades after the end of the war. Part of the explanation lies with the effectiveness of the helicopter, but part also lies with the fact that the helicopter was grafted onto one of the Army's combat branches, a new way to accomplish traditional missions rather than something novel.

The Navy and Riverine Warfare

Whereas the Army adapted the airmobile division, developed for the nuclear battlefield, to the needs of the Vietnam War, the Navy developed a new niche capability—riverine warfare—to wage the war. Because riverine warfare was an adaptation to the particular features of the war, it did not endure; the "brown water" navy became at best a marginal part of the sea service in the years following the war.

The Navy conducted three distinct campaigns in Vietnam: the air war over North Vietnam, maritime operations in the South China Sea, and the brown water patrols conducted throughout South Vietnam's rivers. Elmo Zumwalt, who served as commander of U.S. Naval Forces, Vietnam and later as chief of naval operations, believed that the dominance of the Navy by aviators—including CNO Admiral Thomas Moorer—led to the neglect of less glamorous but more important missions such as coastal patrol and riverine warfare. As he put it, "Air strikes meant glory for the

Navy. He [Moorer] did not want to waste the Navy's resources fighting the war inside Vietnam."[60]

Despite the sea service's strong preference for carrier aviation and blue-water surface operations, the requirements of warfare in Vietnam forced the Navy to develop a niche capability to operate along rivers. As Anthony Harrigan put it, "Orthodoxy has an important place in a naval service. Undue emphasis can be placed on fringe weapons for specialized situations. But the unorthodox also is necessary at times, and now seems one of those times."[61]

The Navy conducted three different coastal and riverine operations, each of which required new technology and doctrine. In 1965, the U.S. 7th Fleet began Operation Market Time, an effort to search and seize enemy supply vessels operating along South Vietnam's 1,200-mile coast. Forces conducing the operation consisted of an air patrol, a surface barrier of large ships, and an inshore patrol of smaller vessels. Although the Navy used a variety of ships for the outer surface barrier, it lacked vessels for inshore operations. To fill this niche, it deployed the Patrol Craft, Fast (PCF), also known as the Swift boat. The fifty-foot aluminum boat was originally designed to transport crews to and from oil platforms in the Gulf of Mexico. It was fast and could travel in very shallow water. Despite the need for many military modifications, the first four boats were delivered to the Navy forty days after they were ordered.[62]

The Navy explored other vessels for Market Time patrols, including patrol gunboats (PGs), hydrofoil patrol gunboats (PGHs), and air-cushioned vehicles. In the fall of 1966, the Navy sent three patrol air cushion vehicles (PACVs) to Vietnam for evaluation. The vehicles, a version of a commercial craft used in England, were capable of seventy knots and had a draft of only one foot. However, they were poorly suited to the patrols, which required neither their high speed nor shallow draft, while their short range and endurance was a major drawback.[63]

Beginning in February 1966, the Navy launched a second riverine operation, Game Warden, to interdict Vietcong infiltration, enforce curfews, and prevent the Vietcong from taxing water traffic. The patrols initially used heavily modified landing craft. Operators jury-rigged spotlights on the boats, added pedestal mounted .50 and .30 caliber guns, and reduced their noise by installing homemade mufflers.[64] With such expedients in hand, almost immediately the Navy began to search for a quieter vessel capable of operating in shallow waters. The result was the patrol boat, river, or PBR. Powered by two 220-horsepower engines with Jacuzzi jets for thrust and steering, the boats could reach a top speed of twenty-eight knots. The vessels were light and could operate in les than a foot of water.

They were also heavily armed, mounting twin .50 caliber machine guns in an open forward mount, a single machine gun on the fantail, and a mount on either side for a Mark 18 grenade launcher or M60 machine gun.[65] The first vessels arrived in Vietnam in March 1966.

The Navy also deployed specialized aircraft to support the riverine patrols. The Army lent the Navy a number of UH-1B Huey helicopters armed with machine guns, 2.75-inch rockets, and grenade launchers. These helicopters, dubbed Seawolves, provided fire support to the PBRs. Beginning in 1969, the Navy augmented the helicopters with twin-engine propeller-driven OV-10A Black Pony aircraft armed with guns, rockets, and flares.[66]

The North Vietnamese adapted to riverine patrols by conducting operations when U.S. forces were not around and employing greater concealment and deception. Still, Game Warden drove many Viet Cong tax collectors off the rivers, disrupted movements, and opened some areas to commercial traffic.[67]

A third innovation was the Mobile Riverine Force, a joint Army-Navy command that used the Navy to transport Army troops on search-and-destroy missions.[68] The force grew out of the need for a mobile force on the Mekong Delta, whose dense population and extensive agriculture made it unsuitable for land basing. It consisted of an assault squadron of Navy ships carrying the 2nd Brigade of the 9th Infantry Division, which was specially configured for the mission. The force used a mobile riverine base that served as billeting and mooring facilities for Army troops and Navy boats. The base included the USS *Benewah* (APB-35) and USS *Colleton* (APB-36), two World War II–era landing ship transports (LSTs) that had undergone extensive conversion, including the addition of jungle green paint, a helicopter platform, air conditioning, and a command center with advanced communications. The USS *Askari* (ARL-30) was activated to provide repair and maintenance capability. A barracks ship, APL-26, provided berthing for 650 sailors and soldiers.[69]

The force employed a range of modified LCM-6 amphibious landing craft. The backbone of the force consisted of armored troop carriers (ATCs) capable of landing a platoon of troops and providing fire support. The craft had a 20mm cannon, two .50 caliber machine guns, and two grenade launchers and were protected by armor plating and metal bars designed to defeat rocket-propelled grenades (RPGs). Other ships were modified as "monitors," the battleships of the riverine force. These were armed with an 81mm mortar and 40mm cannon mounted in a turret. Flamethrowers were added to some to allow them to burn away vegetation and attack bunkers. Other craft were modified into command and control boats (CCBs) equipped with banks of radios to serve as an Army battalion command post.[70] Still others were converted as hospital ships, fitted with flight decks for helicopters, or equipped with fuel bladders.

Riverine operations received a boost after Elmo Zumwalt arrived in Vietnam in October 1968. In November he combined the assets of Market Time, Game Warden, and the Mobile Riverine Force into a deltawide interdiction program, dubbed the South-East Asia Lake, Ocean, River and Delta Strategy (SEA LORDS). He ordered the river patrols to spread out, be more aggressive, and launch hit-and-run attacks into Vietcong territory. Such an approach had a real effect. Among other things, U.S. forces took control of the Rung Sat Special Zone and secured control of the forty-five-mile Long Tau shipping canal to Saigon.[71]

Zumwalt carried his enthusiasm for small craft with him during his tenure as Chief of Naval Operations from 1970 to 1974. As he saw it, the Navy was divided into three powerful "unions": the aviation, submarine, and surface communities. In his view, the fact that his three predecessors had been aviators meant that the service's aircraft needs had received plenty of attention. The submarine community, for its part, had an influential advocate in Admiral Hyman Rickover. The Navy's surface force, by contrast, had been neglected. Zumwalt also felt that activities outside the three "unions," such as mine warfare, surveillance, and communications, lacked a constituency.[72]

In Zumwalt's view, the Navy fielded too much sophisticated (and expensive) weaponry.[73] As CNO, he proposed a "high-low" mix of surface craft: a relatively small number of large, sophisticated ships and a large number of smaller and cheaper vessels capable of performing a wide range of missions. Zumwalt's Project 60 envisioned a naval force that included a class of patrol frigate designed to be half the size and cost of a destroyer armed with a helicopter for antisubmarine warfare (ASW) and Harpoon antiship missiles. The plan also recommended the development of a 17,000-ton, 25-knot sea control ship (SCS). A smaller, cheaper competitor to the nuclear aircraft carrier, the SCS would carry fourteen helicopters and three vertical or short takeoff and landing aircraft. It also envisioned smaller combatants, including a 60-knot hydrofoil armed with *Harpoon* missiles and a surface-effect ship that would skim just above the surface of the ocean at 80–100 knots.[74]

Zumwalt had little success in getting the Navy to adopt such craft. In part, resistance from the Navy's "unions," including his own community of surface warfare officers, undermined his effort. More importantly, it was unclear how such a high-low mix might fit into U.S. strategy. The main argument against less sophisticated ships was that they had limited size and endurance and would be unable to compete with the Soviet navy in time of war. The arguments proved persuasive: only the patrol frigate was ever built, in the form of the *Oliver Hazard Perry*–class frigates.

Three Cases of Innovation

If the United States fought Vietnam largely with forces designed for another type of conflict, it also innovated in several significant cases. In attempting to interdict communist infiltration into South Vietnam, the United States deployed a network of sensors and attack aircraft linked to a command and control center, a precursor to what later analysts would term a "reconnaissance-strike complex." Vietnam also saw the first large-scale use of UAVs and precision-guided munitions. The battlefield debut of these systems only hinted at their potential effectiveness, however. It would only be in the wars of the 1990s and beyond that their value would truly become apparent.

The McNamara Line

One of the central tasks U.S. forces in Vietnam faced was interdicting the flow of men and materiel from North Vietnam to South Vietnam along the so-called Ho Chi Minh Trail—in fact a network of hundreds of roads over ten thousand miles in length.[75] The primary effort to interdict the trail, the so-called McNamara Line, was one of the most prominent efforts to deploy U.S. technology in support of the war effort. It was also an attempt to use technology to solve what was essentially a political problem: the fact that the Vietnamese communists were able to use supposedly neutral Laos and Cambodia to supply their forces in the south. Although it was ultimately a strategic failure, in the process the United States developed and deployed technologies, organizations, and concepts that were in many ways ahead of their time.

The origin of the barrier strategy lay in Secretary of Defense McNamara's disenchantment with strategic bombing, which he saw as ineffective and even counterproductive. In March 1966, he asked the Joint Chiefs of Staff to assess the effectiveness and feasibility of "an iron-curtain counterinfiltration barrier across northern South Vietnam and Laos from the South China sea to Thailand."[76] The military's concept envisioned a five-hundred-yard cleared path surrounding barbed-wire entanglements and an electrified fence, sowed with mines, and observed by bunkers and watchtowers. The JCS estimated that building such a barrier would require 271 battalion-months of engineering, 206,000 tons of construction material, and two to four years to complete. The military was understandably unenthusiastic about such a strategy. General Wheeler, General Westmoreland, and Admiral U.S.G. Sharp, the commander of U.S. Pacific Command, all opposed the plan, fearing it would take troops away from offensive operations.[77]

If a manpower-intensive anti-infiltration barrier was unappealing, McNamara soon had an alternative: a high-technology barrier system. In the summer of 1966, a group of forty-seven civilian scientists convened to consider such a strategy under the aegis of the Jason Division of the Institute for Defense Analysis. By August, they had prepared several papers for McNamara. The group proposed a barrier composed of a thirty-kilometer manned fence stretching from the South China Sea along the southern edge of the Demilitarized Zone to the Annamite Mountains. From there, the barrier would be composed of wide "denial fields" made up of mines and sensors. One paper, entitled "An Air Supported Anti-Infiltration Barrier," envisioned deploying a defensive system that would use small but lethal "gravel mines," a large number of simple sensors, and air strikes to protect South Vietnam. The group also suggested modifying Navy sonobuoys to act as sensors for the barrier.[78] They estimated that such a barrier could be emplaced within a year and maintained for $800 million per year.[79]

McNamara, never one to shrink in the face of military opposition, ordered the project started immediately.[80] On September 15, 1966, he established the highly secret Joint Task Force 728, also known as the Defense Communications Planning Group (DCPG), under Lieutenant General Alfred D. Starbird to head the project and ordered the barrier emplaced within a year.[81] On January 13, 1967, National Security Action Memorandum 358, signed by Walt Rostow, gave the barrier the highest priority for funding.[82]

The barrier had historical precedents. In the 1950s the French constructed the Morice Line to prevent rebel infiltration from Tunisia into Algeria. The barrier consisted of barbed wire, an electrified fence, and antipersonnel mines. Searchlights, roving patrols, and electronic sensors reinforced the barrier. The strategy was a tactical success, reducing infiltration 90 percent, but failed to deliver a strategic victory: France relinquished control of Algeria in 1962.

The U.S. anti-infiltration barrier was composed of sensors, relay aircraft, and a fusion center, the Infiltration Surveillance Center (ISC) at Nakhon Phanom in Thailand. It was a highly classified program that operated under a variety of code names throughout its life. For example, the project was originally given the code name Project Nine. After that was compromised, it was given the name Dye Marker. The air portion of the barrier was named Muscle Shoals and associated technologies Igloo White.

One of the greatest challenges was developing sensors to detect North Vietnamese infiltration. The Defense Department spent $670 million to field a large inventory of sensors that could be hidden in the thick vegetation of the region.[83] Some used thermal, electromagnetic, or chemical sensors to detect engine or body heat, electrical or magnetic field fluctuations, machinery noises, or even smell.[84] Some had spikes to implant themselves in the soil; others were designed to hang from the jungle canopy. The U.S. Navy adapted airdropped radio sonobuoys for ground use by replacing their hydrophones with microphones and seismic sensors and modified antisubmarine aircraft for use over land. In fact, in many ways the McNamara Line was a land-based application of the Navy's approach to antisubmarine warfare.

The most widely deployed sensors were seismic. The first to be introduced in large quantities was the Seismic Intrusion Detector, which arrived in Vietnam in October 1967. Although it was put to good use by various artillery firebases, Special Forces, and the 1st Air Cavalry Division, it suffered from a high false-alarm rate. The most common variety was the Air Delivered Seismic Intrusion Detector (ADSID); 36,000 were produced by one manufacturer alone. Another variety, the Ground Seismic Intrusion Detector (GSID), was a seven-pound sensor packed into a box the size of a brick. Unlike the ADSID, the GSID was designed to be emplaced by hand.[85] The Patrol Seismic Intrusion Detector, or PSID, was a set of four sensors and a receiver small enough to be carried and used by squad-size patrols.[86] Although such sensors proved to be the most reliable variety deployed in Vietnam, that was a dubious distinction: they were plagued by numerous false alarms caused by tremors, wind, thunder, rain, nearby bomb and artillery detonations, and passing aircraft.[87]

In order to locate activity along the Ho Chi Minh Trail, it was important to have an accurate record of the sensors' locations. This proved to be a major challenge. The Navy initially modified four OP-2E Neptune ASW aircraft to deliver sensors at very slow speeds from as low as five hundred feet. The aircraft gained a tail gunner with a night observation scope and a 20mm cannon, as well as an AN/APQ-29 search radar, forward-looking infrared (FLIR) sensor, side-looking airborne radar, and low-light level television (LLLTV). However, the slow, low-flying aircraft made easy targets for North Vietnamese gunners and enjoyed only limited accuracy in delivering sensors. Later, Air Force Special Operations forces dropped sensors by hand from CH-3 Jolly Green Giant helicopters. These offered better accuracy but remained vulnerable to ground fire. Eventually the Air Force began using long-range navigation (LORAN)-equipped F-4Ds to drop the sensors.[88] Others were emplaced by hand by South Vietnamese Special Forces Spike teams.

The sensors transmitted their data to relay aircraft orbiting overhead and on to the ISC. Lockheed EC-121R radio relay aircraft flew in four constantly manned aerial orbits over Southeast Asia. These aircraft, equipped with thirteen multichannel communications transceivers, operated under the call sign Batcat. Radio relay aircraft initially had difficulty picking up the sensors' transmissions. In 1968, it was estimated that they were able to monitor only 40–60 percent of transmissions. By 1969, however, they were monitoring more than 80 percent.[89]

The Batcat aircraft were augmented by a joint Air Force/DARPA program entitled Pave Eagle, which modified six Beech A-36 Debonaire aircraft, renamed the QU-22B, as radio relay aircraft. Although the QU-22B was to be an unmanned aircraft, all operational flights carried a pilot due to concerns about reliability. The program was cancelled in 1972 after the loss of several aircraft.

The heart of the system was the ISC at Nakhon Phanom in northeast Thailand. The largest single building in Southeast Asia, the ISC was staffed by four hundred Americans under the command of General William Mc-Bride. At the ISC, 7th Air Force analysts analyzed raw sensor data, determined the nature of the activity, and dispatched aircraft to strike targets.[90] They monitored the situation along the Ho Chi Minh Trail complex on a twenty-four-foot high, nine-foot wide Plexiglas map. They used two huge IBM 360/Model 65 computers to analyze sensor readings and transform them into targeting information. As one Air Force officer put it, "We got the Ho Chi Minh Trail wired like a pin ball machine."[91]

It was not enough to receive and process sensor data: forward air controllers needed to identify targets and vector strike aircraft to them. The Air Force devised a number of innovations to strike targets along the trail. To allow aircraft to operate at night and in all weather, the Air Force refitted a number of its F-4 and RF-4 aircraft with LORAN navigation equipment. The 7th Air Force also modified several Combat Skyspot radar sites to allow forward-air controllers to guide fighters to targets along the trail. The Air Force also upgraded eleven B-57G Canberra bombers with forward-looking radar, infrared sensors, low-light television and laser ranging devices for operations along the trail.[92] Finally, it fielded the Battlefield Illumination Airborne System (BIAS), which equipped C-123 cargo aircraft with xenon arc lamps coupled with downward-looking infrared and forward-looking radar. From 12,000 feet the device could illuminate a circle two miles in diameter with four times the illumination of a full moon.[93]

The U.S. government explored technologies that were not only innovative, but also downright bizarre. ARPA, in 1972 renamed the Defense Advanced Research Projects Agency (DARPA), conducted Project Agile, which included an attempt to produce a mechanical elephant designed to traverse on servomechanism "legs" through jungle too thick for jeeps.[94] Scientists explored the use of silver iodide to seed clouds to flood the Ho Chi Minh Trail. Beginning in May 1967, U.S. forces commenced an operation designed to turn the trail complex into mud. The operation, code named Commando Lava, included air crews dropping sacks of a mud-making compound from C-130 aircraft. However, the effort had little effect; the Vietnamese covered muddy areas with logs and mats and went on their way.[95]

Construction of the McNamara Line began in May 1967, when U.S. engineers began clearing the initial trace of the barrier. In subsequent months, McNamara visited Vietnam twice. He was impressed with work on the barrier and revealed its existence upon returning to the United States on September 8, 1967.[96] JTF 728 ordered more than five million fence posts and roughly fifty thousand miles of barbed wire. The barrier would cost $3 billion to $5 billion overall, require the use of twenty-three thousand acres of land, and lead to the resettlement of between thirteen thousand and eighteen thousand villagers. In December 1967, the antipersonnel array around the DMZ and Tchepone, code-named Dump Truck, became operational. In January 1968 the antivehicular array, code-named Mud River, became active.[97]

The barrier was a case study in the American way of war. It was a massive undertaking, the product of a nation with enormous industrial and intellectual resources. It nonetheless got off to a slow start. Although analysts at ISC identified thirty-eight targets during its first week in operation, only four aircraft found targets and two struck them.[98] The poor performance was the result of the fact that other high priority targets were available, as well as the fact that visual confirmation of a target was required before attack.

Antivehicle barrier operations, christened Commando Hunt, accounted for hundreds of heavy bomber, fighter-bomber, and gunship sorties each year between 1968 and 1972. Assessing the effectiveness of these operations, however, is difficult. By Commando Hunt V in 1971, twenty-one thousand trucks had been reported destroyed or damaged.[99] However, often the only bomb damage assessment that the Air Force received was from pilot reports, which consistently overestimated kills.

The North Vietnamese developed countermeasures to the U.S. sensors. They constructed covered roads through the jungle to shield them from overhead observation. Drivers coordinated their operations so that they would operate during periods of the least aircraft activity.[100] The North Vietnamese also found other ways to deliver supplies to South Vietnam by sea through, for example, Cambodia. Because the insurgency in the south required minimal logistical support, it was easy to keep it going. Moreover, the interdiction campaign could not prevent North Vietnam from intervening in force, as it did during the 1972 Easter Offensive.

Perhaps the most successful use of ground sensors was unanticipated by the architects of the barrier strategy. U.S. forces were installing sensors in conjunction with the barrier when the marine base at Khe Sanh was attacked in January 1968. Westmoreland diverted the sensors from the barrier to ring the base. Local commanders later testified that twice as many marines would have been killed in the subsequent siege if the sensors had not been deployed around the base's perimeter.[101] So successful did the sensors prove at warning of enemy movement and identifying targets that Westmoreland obtained permission to delay the completion of the McNamara Line in order to use the sensors in tactical operations.

In April 1968, the U.S. military launched Operation Duffel Bag, which involved the use of unattended ground sensors for other missions, such as ambushes, base defense, and monitoring landing zones. Duffel Bag received nine times as many sensors as the McNamara Line, and while they were active only one-fourth as often as those on the McNamara Line, they were three times as likely to lead to air strikes.[102] By 1969, MACV had installed sensors on the perimeters of military installations, along main convoy routes, and across principal enemy avenues of approach.[103]

In 1971, the U.S. military launched Mystic Mission, a European version of the McNamara Line. The U.S. military conducted demonstrations of unmanned sensor technologies at Eglin AFB in Florida in 1971 and in Germany in 1972.[104] By March 1974, the director of defense research and engineering, Malcolm R. Currie, was willing to testify that "A remarkable series of technical developments has brought us to the threshold of what I believe will become a true revolution in conventional warfare."[105]

To some, the McNamara Line heralded the arrival of what later theorists would dub a "reconnaissance-strike complex." As Westmoreland predicted in a speech to the Association of the U.S. Army in October 1969, "I see battlefields or combat areas that are under 24-hour real or near-real-time surveillance of all types. I see battlefields on which we can destroy anything we can locate through instant communications and the almost instantaneous application of highly lethal firepower."[106]

Despite such lofty rhetoric, many of the underlying technologies, pressed into service in Vietnam, remained immature. It was not until the 1990s, with further developments in sensor, communication, and data-processing technology that they would begin to display their effectiveness.

Remotely Piloted Vehicles

Vietnam also witnessed the first extensive use of early generation UAVs. The Air Force flew some 3,500 UAV sorties during the war. However, the Air Force did not exploit the success of unmanned systems in the years that followed the war's end. Indeed, UAVs did not gain acceptance until the 1990s.[107]

The most common UAV was the Teledyne Ryan BQM-34 Firebee, also known as the Lightning Bug. The drone was a development of the BQM-24A jet-powered target drone. The Firebee flew its first mission over Vietnam and China in August 1964. Throughout the war, the U.S. Air Force launched some 2,350 of the drones. One-third the size and one-twentieth the weight of a fighter, the jet-powered drone could fly at high subsonic speeds. While the standard reconnaissance version carried a high-resolution camera, its versatile design could accommodate numerous modifications and a variety of payloads. From an altitude of 1,500 feet it could photograph a swath 120 nautical miles long and three nautical miles wide with a resolution of one foot.[108]

Although the drones themselves were much less expensive than manned aircraft, their flight operations were elaborate. A drone was launched from a DC-130 mothership and flew a preplanned track. After conducting its photo run, the drone would climb, cut off its engines, and deploy parachutes. Support personnel would attempt to recover the drone by helicopter; if not, they would recover it on the ground.[109] Nearly thirty personnel were involved in each mission, and each drone was able to conduct less than one flight per day.[110] Another significant limitation was the relative inaccuracy of its navigational system. Drones took photographs of less than half of their planned reconnaissance targets, mainly due to navigation errors.[111]

In addition to being a significant source of reconnaissance imagery and bomb damage assessment, UAVs were used in a number of innovative roles during the war. In February 1966, the Air Force conducted Project United Effort, which used a specially configured drone to act as bait for North Vietnamese SAMs. The CIA developed a special electronics package for the Firebee that would record the missile's transmissions as it homed in on its target, fused, and exploded. The data, beamed to an aircraft orbiting nearby, allowed U.S. electronics experts to build countermeasures to the missile. Later that year, another specially converted drone flew with a Navy jamming pod to test its effectiveness. The drone drew more than ten North Vietnamese SAMs before being destroyed, demonstrating the effectiveness of U.S. electronic countermeasures.[112]

While UAVs saw considerable use in the war, they did not find a permanent home in the Air Force until decades later. The Air Force pursued three UAV programs during the 1970s—Compass Dwell, Compass Cope, and a modified Lightning Bug—but none reached maturity. In each case, cost, the availability of manned substitutes, arms control constraints, and air space limitations conspired to make unmanned systems unattractive. Favored by neither the bomber nor the fighter communities, unmanned systems lacked an organizational home. It was not until the 1990s that UAVs came into their own.[113]

Precision-Guided Munitions

The war also saw the first widespread use of laser-guided bombs (LGBs). Although these precision-guided munitions (PGMs) had a significant tactical impact, they were introduced too late to have more of an effect.

The Navy and Air Force's primary guided air-to-surface missile at the beginning of the Vietnam War was the radio-controlled, gyro-stabilized Bullpup. The missile was fielded in two versions: a "Little" Bullpup with a 250-pound warhead and a "Big" Bullpup with a 1,000-pound warhead. Although the weapon offered improved accuracy over unguided weapons, it suffered from a number of shortcomings. Because it had to be directed to its target by an operator throughout its flight, only one missile could be guided at a time. An aircraft therefore needed to make multiple passes to expend its load, increasing its vulnerability. The weapon also suffered from poor reliability. Even more disconcerting, its warhead proved too small; even when it struck its target, in some cases it bounced off the bridges that were its targets.[114]

As a result of these shortcomings, the Navy and Air Force developed the Walleye, a free-fall bomb with an 829-pound warhead guided by an electro-optical sensor. Once an operator locked onto his target and released the missile, it would home in autonomously. The weapon proved successful when operating in good weather against highly visible and lightly defended targets, but performed poorly under more demanding conditions.[115] Moreover, it was limited to low-level delivery, a tactic that exposed aircraft to hostile fire.

LGBs were one of the most significant innovations of the Vietnam War. The service that initially had the greatest interest in military applications for the laser was the U.S. Army. The Army's Missile Command used laser guidance for indirect-fire and antitank weapons. Army engineers hoped to use a laser to illuminate a target and then design a seeker system to guide a missile to it. In June 1963, Missile Command contracted with North American-Autonetics (NA-A) and RCA-Burlington to develop laser seekers. By the end of the following year, both had demonstrated laser guidance systems in the laboratory.[116]

The Army's interest in laser-guided weapons waned as its involvement in Vietnam escalated. Laser-guided antitank weapons appeared to have little relevance to the problems the Army faced in Southeast Asia, and in 1965 the service reduced its funding for laser research. However, when Missile Command decided to offer the results of its research to other services, the Air Force eagerly stepped forward to continue the work.[117]

Under Air Force sponsorship, NA-A and Texas Instruments (TI) pursued two different approaches to laser guidance: NA-A's was more complex but also more feasible, while TI's was less complex but unproven. In addition, TI's prototype was one-third the cost of NA-A's.[118] The Air Force eventually awarded TI the contract.

In March 1967 the Air Force issued an operational requirement for precision bombing, driven by the need to increase bombing effectiveness. The development of laser-guided bombs, dubbed Project Paveway, was assigned the highest funding priority. Six months later, the Air Force established a formal requirement for bomb with a CEP of no greater than twenty-five feet and a guidance reliability of 80 percent.

The speed with which the first LGBs were developed was remarkable: the requirement for a LGB was established nine months after the feasibility of laser guidance had been established, two years after prototype contracts had been let, and two and a half years after Missile Command had first briefed the Air Force on the feasibility of laser guidance.[119]

On January 15, 1968, the Air Force approved the purchase of 293 seeker kits at a cost of approximately $16,000 per kit. These were attached to Mark 84 2,000-pound bombs. A small number of these weapons were tested at Eglin AFB, Florida, before being sent to Southeast Asia between May and August 1968. The encouraging results of these tests led the Air Force to purchase an additional thousand laser kits.[120]

LGBs saw extensive use in Vietnam. The Navy and Air Force expended more than 28,000 LGBs in Southeast Asia between 1968 and 1973, mainly against bridges and transportation chokepoints.[121] The Paveway I in particular proved highly successful. During 1969, Air Force crews delivered 1,601 2,000-pound Paveways, 61 percent of which scored direct hits.[122] The weapons were also inexpensive: in 1968 the estimated cost of a 750-

pound bomb with seeker was $7,149 and of a 2,000-pound bomb was $7,930. Moreover, the combination of economies of scale and engineering improvements lowered the cost of the seeker dramatically. Between 1968 and 1971, the initial cost of the seeker fell from $11,800 to $4,100.[123]

The introduction of LGBs had a dramatic effect on the bombing campaign over North Vietnam. Between February 1972 and February 1973, the Air Force expended more than 10,500 LGBs, mostly 2,000-pound weapons. Nearly 50 percent were assessed as direct hits, while another forty percent achieved a CEP of twenty-five feet.[124] The bombs' accuracy was thirty-three to fifty times better than F-105 dive-bombing with unguided weapons during Rolling Thunder. In that campaign, F-105s averaged a CEP of around five hundred feet. The campaign pointed to a future in which air-to-ground strike operations would be built around guided munitions that could reliably land within ten to twenty feet of their targets.

The deployment of aircraft armed with LGBs and equipped with laser designators affected the composition of strike formations. During Rolling Thunder, the standard strike package was composed of thirty-six aircraft, all of equal value. During Linebacker, the strike component was eight to twelve more valuable aircraft. Because there were relatively few Pave Knife aircraft in theater, the Air Force went to great lengths to protect them. As a result, strike packages became larger and more complex. Because LGB effectiveness depended on clear weather, weather aircraft preceded Linebacker missions. Next over the target was a MiG Combat Air Patrol of two to four aircraft, ideally including two Combat Tree–equipped F-4s. These were followed by eight to twelve F-4s armed with Mk-129 chaff bombs. These aircraft laid a chaff corridor to degrade the performance of SAM and AAA radars. Eight F-4s equipped with ECM pods provided a close escort of the LGB-equipped strike formation. In addition, the formation would use EB-66s for standoff jamming and Wild Weasels for SAM suppression.[125]

LGBs saw widespread use during the 1972 Easter Offensive, during which the Air Force used the weapons against North Vietnamese forces and supporting infrastructure, such as the bridges that North Vietnamese armored forces needed to cross rivers. Perhaps the best-known case is that of the Thanh Hoa Bridge. By May 1972, 871 Navy and Air Force sorties had flown against the bridge, including an attempt to destroy it by having a C-130 Hercules transport plane deliver huge floating mines against it. All failed, and eleven aircraft were shot down in the process. On April 27, 1972, aircraft carrying several 2,000-pound bombs with electro-optical guidance attacked the bridge. The bridge was damaged but remained standing. On May 13, it was attacked with 2,000- and 3,000-pound LGBs that dropped one of the main spans and damaged others, rendering the bridge unusable. Such success was repeated over and over. Less than a month after the resumption of bombing in the north, thirteen key bridges along the rail lines linking North Vietnam and China were dropped. Between April 6 and the end of June, the 8th Tactical Fighter Wing (TFW) alone destroyed 106 bridges.[126]

LGBs also permitted the United States to strike targets located near residential areas, cultural landmarks, and other sensitive sites. While LGBs provided improved accuracy, they were by no means perfect: some weapons still went astray, causing civilian casualties and giving the North Vietnamese a propaganda weapon. Hanoi skillfully manipulated such accidents to allege that the United States was deliberately bombing noncombatants.[127]

The Vietnam War was a peripheral campaign in the Cold War. It is thus hardly surprising that the services relied heavily upon formations organized, trained, and equipped to fight the Soviets to wage war in Southeast Asia. Such a formulation, however, ignores the fact that the conflict also witnessed significant innovation. Heavy reliance on (if not development of) airmobile forces, the formation of riverine units, the development of fixed-wing gunships, and the tactical employment of strategic bombers were all responses to the unique features of the war. Moreover, the United States also deployed for the first time a variety of new weapons, such as unattended ground sensors, UAVs, and LGBs, whose impact would far outlast the war. Between 1991 and 2003, for example, roughly half of U.S. PGM expenditures were LGBs.[128]

The structure and culture of the services affected how new ways of war were received. Those approaches that could appeal to an existing constituency within the service, such as the formulation of airmobile operations as a new form of cavalry, flourished. Approaches that complemented a service's identity similarly flourished. For example, LGBs didn't threaten Air Force or Navy aviators; all they did was make existing tasks more effective.

By contrast, approaches to warfare that did not resonate with existing cultures within the services remained marginal. The pilots of Air Force gunships never gained the type of prestige or power that those of bombers, attack aircraft, or fighters had. Similarly, riverine operations fit poorly with the culture of the Navy's surface community. Both migrated to their services' special operations communities and, eventually, to U.S. Special Operations Command.

America's reliance on technology was hardly the cause of its defeat in Vietnam. Nor could it have been a source of victory. The United States lost the war because of an inability to develop a strategy to achieve its aim of a free, independent, noncommunist South Vietnam.

Notes

1. See, for example, Andrew F. Krepinevich Jr., *The Army and Vietnam* (Baltimore: Johns Hopkins University Press, 1986); Earl H. Tilford Jr., *Setup: What the Air Force Did in Vietnam and Why* (Maxwell AFB, AL: Air University Press, 1991).
2. Russell F. Weigley, *The American Way of War: A History of United States Military Strategy and Policy* (Bloomington: Indiana University Press, 1973), 465–66.
3. Kenneth P. Werrell, *Chasing the Silver Bullet: U.S. Air Force Weapons Development from Vietnam to Desert Storm* (Washington, DC: Smithsonian Books, 2003), 13.
4. Ibid., 42.
5. Marshall L. Michel III, *Clashes: Air Combat Over North Vietnam, 1965–1972* (Annapolis, MD: Naval Institute Press, 1997), 19.
6. Ibid., 29.
7. Werrell, *Chasing the Silver Bullet*, 49.
8. Gilles Van Nederveen, *Sparks Over Vietnam: The EB-66 and the Early Struggle of Tactical Electronic Warfare* (Maxwell AFB, AL: Air University Press, 2000), 11–13, 44–45.
9. Michel, *Clashes*, 37, 61, 71, 38.
10. Quoted in ibid., 72.
11. Ibid., 34.
12. Ibid., 33.
13. Ibid., 35.
14. Werrell, *Chasing the Silver Bullet*, 50.
15. Michel, *Clashes*, 62.
16. Ibid., 41, 45.
17. Ibid., 46.
18. Ibid., 100.
19. Ibid.
20. Werrell, *Chasing the Silver Bullet*, 48.
21. Michel, *Clashes*, 114.
22. Ibid., 252.
23. Ibid., 44–45, 151.
24. Curtis Peebles, *Dark Eagles: A History of Top Secret U.S. Aircraft Programs* (Novato, CA: Presidio Press, 1995), 219.
25. Werrell, *Chasing the Silver Bullet*, 43.
26. Michel, *Clashes*, 154; Werrell, *Chasing the Silver Bullet*, 45.
27. Werrell, *Chasing the Silver Bullet*, 45.
28. Michel, *Clashes*, 228–30.
29. Peebles, *Dark Eagles*, 218–19.
30. Michel, *Clashes*, 181, 267, 268.
31. Ibid., 186.
32. Peebles, *Dark Eagles*, 219–20, 223–24.

33. Worden, *Rise of the Fighter Generals*, 158.
34. Ibid., 173.
35. Van Nederveen, *Sparks Over Vietnam*, 35.
36. Werrell, *Chasing the Silver Bullet*, 31.
37. Ibid., 32.
38. Timothy N. Castle, *One Day Too Long: Top Secret Site 85 and the Bombing of North Vietnam* (New York: Columbia University Press, 1999), 1–2, 15.
39. Edward F. Puchalla, "Communist Defense Against Aerial Surveillance in Southeast Asia," *Studies in Intelligence* 14, no. 2 (Fall 1970): 31–78.
40. Werrell, *Chasing the Silver Bullet*, 18.
41. Ibid., 19.
42. Ibid., 20–21.
43. Worden, *Rise of the Fighter Generals*, 172, 186.
44. Krepinevich, *The Army and Vietnam*, 4–7.
45. John H. Hay, *Tactical and Materiel Innovations: Vietnam Studies* (Washington, DC: U.S. Government Printing Office, 1974), 179.
46. Krepinevich, *The Army and Vietnam*, 198.
47. James M. Gavin, "Cavalry, and I Don't Mean Horses," *Harper's Magazine*, April 1954, 55.
48. J. A. Stockfisch, *The 1962 Howze Board and Army Combat Developments* (Santa Monica, CA: Rand Corporation, 1994), 9–10.
49. Department of Defense, *Semiannual Reports, Jan.–June, 1956*, 83.
50. Robert S. McNamara, "Memorandum for Mr. Stahr," April 19, 1962, reprinted in Stockfisch, *1962 Howze Board*, 41.
51. Stockfisch, *1962 Howze Board*, 1.
52. Ibid., 24.
53. Ibid., 21, 24.
54. Ibid., 25–26.
55. Matthew Allen, *Military Helicopter Doctrines of the Major Powers, 1945–1992: Making Decisions About Air-Land Warfare* (Westport, CT: Greenwood, 1993), 10.
56. Stockfisch, *1962 Howze Board*, 28.
57. Jones, *Arming the Eagle*, 381.
58. Hay, *Vietnam Studies*, 82.
59. Quoted in John D. Bergen, *Military Communications: A Test for Technology* (Washington, DC: U.S. Government Printing Office, 1989), 284.
60. Elmo Zumwalt Jr., Elmo Zumwalt III, and John Pekkanen, *My Father, My Son* (New York: Macmillan, 1986), 44.
61. Quoted in Jones, *Arming the Eagle*, 386–88.
62. Jones, *Arming the Eagle*, 390.
63. Thomas J. Cutler, *Brown Water, Black Berets: Coastal and Riverine Warfare in Vietnam* (Annapolis, MD: Naval Institute Press, 1988), 91.
64. Ibid., 143–44.
65. Jones, *Arming the Eagle*, 388.
66. Cutler, *Brown Water*, 193, 195–97.
67. Baer, *One Hundred Years of Sea Power*, 390.
68. William B. Fulton, *Riverine Operations: 1966–1969* (Washington, DC: Department of the Army, 1973).
69. Cutler, *Brown Water*, 248–49.

70. Ibid., 240–44.
71. Baer, *One Hundred Years of Sea Power*, 391–92.
72. Elmo R. Zumwalt Jr., *On Watch: A Memoir* (New York: Quadrangle, 1976), 64.
73. Ibid., 72.
74. Baer, *One Hundred Years of Sea Power*, 405.
75. Darrel D. Whitcomb, "Tonnage and Technology: Air Power on the Ho Chi Minh Trail," in *Military Aspects of the Vietnam Conflict*, ed. Walter L. Hixson (New York: Garland, 2000), 236.
76. John Prados, *The Blood Road: The Ho Chi Minh Trail and the Vietnam War* (New York: Wiley, 1999), 213.
77. Ibid.
78. Christopher P. Twomey, "The McNamara Line and the Turning Point for Civilian Scientist-Advisers in American Defense Policy, 1966–1968," *Minerva* 37 (1999): 242–44.
79. Prados, *The Blood Road*, 214.
80. Twomey, "The McNamara Line," 244.
81. Bergen, *Military Communications*, 392; Prados, *The Blood Road*, 213.
82. Prados, *The Blood Road*, 214.
83. Bergen, *Military Communications*, 392.
84. For example, one device, the "people sniffer," was an airborne electrochemical instrument that sensed microscopic particles suspended in the air. Mounted on helicopters, the sniffer sampled the atmosphere for evidence of the enemy. See Hay, *Vietnam Studies*, 80.
85. "Acoubuoy, Spikebuoy, Muscle Shoals, and Igloo White," at http://home.att.net/~c.jeppson/igloo_white.html.
86. Twomey, "The McNamara Line," 251.
87. Werrell, *Chasing the Silver Bullet*, 37–38.
88. Ibid., 38.
89. Ibid., 39.
90. Bergen, *Military Communications*, 392.
91. Quoted in Whitcomb, "Tonnage and Technology," 239.
92. Ibid.
93. Werrell, *Chasing the Silver Bullet*, 40.
94. Charles Piller, "Army of Extreme Thinkers," *Los Angeles Times*, August 14, 2003.
95. Prados, *The Blood Road*, 220–21.
96. Twomey, "The McNamara Line," 247.
97. Prados, *The Blood Road*, 219.
98. Twomey, "The McNamara Line," 248.
99. Whitcomb, "Tonnage and Technology," 240.
100. Ibid., 242.
101. Twomey, "The McNamara Line," 249.
102. Ibid., 251.
103. Bergen, *Military Communications*, 392.
104. Paul Dickson, *The Electronic Battlefield* (Bloomington: Indiana University Press, 1976), 122–23.
105. Quoted in ibid., 159.
106. Ibid., 221.

107. Thomas P. Ehrhard, "Unmanned Aerial Vehicles in the United States Armed Services: A Comparative Study of Weapon System Innovation" (Ph.D. dissertation: Johns Hopkins University, 2000), 50.

108. Werrell, *Chasing the Silver Bullet*, 33.

109. Ibid., 35.

110. Ehrhard, "Unmanned Aerial Vehicles," 408.

111. Ibid., 408.

112. Ibid., 412–14.

113. Ibid., 438–64.

114. Werrell, *Chasing the Silver Bullet*, 143.

115. Ibid.

116. Peter deLeon, *The Laser-Guided Bomb: Case History of a Development*, R-1312-1-PR (Santa Monica, CA: Rand Corporation, 1974), 6.

117. Ibid., 7.

118. Ibid., 12.

119. Ibid., 20–23.

120. Ibid., 23–26.

121. Headquarters, U.S. Air Force, Management Information Division, *United States Air Force Statistical Digest: Fiscal Year 1973*, July 31, 1974, table 34, p. 86; *United States Air Force Statistical Digest: Fiscal Year 1974*, April 15, 1975, table 37, p. 73.

122. Werrell, *Chasing the Silver Bullet*, 149.

123. David R. Mets, *The Quest for a Surgical Strike: The United States Air Force and Laser Guided Bombs* (Eglin AFB, FL: Office of History, Armament Division, Air Force Systems Command, 1987), 71, 94–95.

124. Donald K. Osterman, "An Analysis of Laser Guided Bombs in SEA," Headquarters 7th Air Force, Thailand, Tactical Analysis Division, Air Operations Report 73/4, June 28, 1973, ii, 9, 34.

125. Michel, *Clashes*, 218–25.

126. Mets, *Quest for a Surgical Strike*, 86–87.

127. Ibid., 91.

128. I am grateful to Barry D. Watts for this information.

4

Winning the Cold War, 1976–1990

IN THE DECADE and a half following the end of the Vietnam War, the U.S. military transformed itself from a defeated force to one that achieved victory in the Cold War. This success had several ingredients, including a challenging adversary and a battle-hardened officer corps that was determined not to repeat the mistakes of the last war. The development and deployment of a new generation of weapons and the doctrine needed to employ them effectively was an important—some would argue the most important—element of the U.S. strategy to compete with and ultimately defeat the Soviet Union. During the Carter and Reagan administrations, technology came to be seen as a key arena of superpower rivalry. U.S. technological superiority was of critical importance in developing ways to counter the Soviet Union's tank armies in Central Europe, its maritime bomber force, and its nuclear missile arsenal.

U.S. Technology in the Late Cold War

It is difficult for those who did not live through the 1970s to understand the decade's zeitgeist. The former chairman of the Joint Chiefs of Staff, Admiral Thomas H. Moorer, captured the mood accurately when he wrote in

1977, "the United States is crossing the threshold of the last quarter of the 20th century in a mood of apprehension and confusion—confusion over America's place in a rapidly-changing world and over the correct path to a dimly perceived future."[1] The Vietnam War sapped the strength and morale of the U.S. military. In the years that followed, the Defense Department had to contend with low budgets and weak public support for the military.

Soviet power, by contrast, appeared ascendant, particularly in Central America, the Caribbean, Africa, and the Middle East. The 1970s saw an increase in the size and the sophistication of the Soviet armed forces. A 1981 CIA assessment of Soviet military developments concluded that over the preceding fifteen years the Soviet Union had increased its stockpile of intercontinental nuclear delivery vehicles nearly sixfold; maintained the world's largest strategic defense and civil defense programs; more than tripled the size of its battlefield nuclear forces; more than doubled its divisions' artillery firepower; increased ninefold the weight of ordnance that its tactical air forces could deliver deep in NATO territory; introduced new surface ships, submarines, and naval aircraft; broadened its military activities in the third world; nearly doubled its defense expenditure in real terms; and more than doubled military research and development investment.[2] An assessment published two years later predicted that the Soviet Union would field a greater number of new or substantially modified weapons in the 1980s than it had in each of the previous two decades. "As a result… Soviet leaders are expected to have available an unprecedented number of weapon systems that can be deployed with military forces through the early 1990s."[3]

Of particular concern was the growing sophistication of Soviet military hardware. The 1984 edition of the Department of Defense publication *Soviet Military Power* concluded, "While the United States continues to lead the USSR in most basic technologies, the gap continues to narrow in the military application of such technologies. Increasingly, the incorporation of critical Western technologies is permitting the USSR to avoid costly R&D efforts and to produce, at a much earlier date than would otherwise be possible, Soviet weapons comparable to or superior to fielded US weapons."[4]

Under both Carter and Reagan, the defense department sought to exploit the U.S. advantage in advanced technology to offset the Soviet Union's numerical superiority. In essence, such a strategy attempted to use the U.S. lead in information technology to counter the Soviet edge in heavy industry.

During the tenure of Jimmy Carter's secretary of defense, Harold Brown, the department developed the so-called offset strategy, which envisioned

using advanced technology to counterbalance the Warsaw Pact's numerical advantage. This strategy made explicit one of the central assumptions
of postwar U.S. defense planning: that the American lead in technology
could give its armed forces a significant battlefield edge. Under Brown,
the defense department sought to use modern electronics and computers to multiply the effectiveness of U.S. forces. It also invested in weapon
systems that it hoped would render the accumulated stockpile of Soviet
equipment obsolescent, such as stealthy aircraft and electronic countermeasures.[5] For example, defense planners looked to a range of Army and
Air Force systems, including precision-guided munitions such as the Copperhead artillery projectile and Hellfire antitank missile, to counter Soviet
tank formations.[6]

Exploiting the U.S. technological edge in the competition with the Soviet Union became explicit policy during the Reagan administration. As
National Security Decision Directive 75, "U.S. Relations with the USSR,"
put it:

> The U.S. must modernize its military forces—both nuclear and conven-
> tional—so that Soviet leaders perceive that the U.S. is determined never
> to accept a second place or a deteriorating military posture. Soviet calcula-
> tions of possible war outcomes under any contingency must always result
> in outcomes so unfavorable to the USSR that there would be no incentive
> for Soviet leaders to initiate an attack. The future strength of U.S. military
> capabilities must be assured. U.S. military technology advances must be
> exploited, while controls over transfer of military related/dual-use technol-
> ogy, products, and services must be tightened.[7]

In so doing, the U.S. government exploited Soviet fears, reported by the
CIA, of being outpaced technologically by the United States.[8]

The Soviets maintained a large effort, both overt and covert, to acquire U.S. technology. A network of KGB and military intelligence agents
targeted U.S. industry, particularly electronics, computers, and manufacturing technology.[9] For example, the Soviets were able to acquire
documentation for the F/A-18 Hornet aircraft that saved them over a
thousand man-years of research. The Hornet's fire-control radar served
as the basis for the look-down/shoot-down radars for the Soviet MiG-29
Fulcrum and Su-27 Flanker. Information stolen from the United States
served as the impetus for Soviet projects to design new radar-guided
air-to-air missiles (AAMs) and improved countermeasures against U.S.
radar systems.[10]

Scientific espionage also offered evidence of Soviet technological
weakness. The United States thus undertook several efforts to shape So-

viet perceptions of the technological competition. One involved feeding deceptive information to the Soviets regarding the state of U.S. military technology. In 1981, French intelligence recruited Colonel Vladimir I. Vetrov, a KGB officer who had been assigned to collect intelligence on Western science and technology. Vetrov, dubbed "Farewell," gave the French a shopping list of the technologies the Soviets were seeking, information the French passed on to the Americans.[11] In early 1984, the CIA and Pentagon used their knowledge of Soviet collection requirements to begin feeding Moscow incomplete and misleading information. The disinformation campaign covered half a dozen sensitive military technologies that the Soviets were interested in, including stealth, ballistic missile defenses, and advanced tactical aircraft. The United States planted false information regarding development schedules, prototype performance, test results, production schedules, and operational performance.[12] The Defense Department also inaugurated a deception program associated with the Strategic Defense Initiative as a way to make the Soviets believe that U.S. ballistic missile defense (BMD) capabilities were more formidable than they in fact were.[13]

The Military Reform Movement

Not all agreed that the U.S. lead in advanced technology was an inherent source of advantage. The 1980s witnessed the emergence of the so-called military reform movement, whose members charged that the military brass favored weapons that cost too much and which were of questionable effectiveness. As one of the movement's more articulate spokesmen, William S. Lind, wrote, "The defense establishment's concept of quality leads to weapons that push the technological state of the art but often do so in areas that have little relevance to actual combat. These weapons also tend to be fragile and very difficult to maintain in the field, often fail to perform under combat conditions—which are very different from conditions on proving grounds—take decades to develop, and are extremely expensive both to buy and operate."[14] James Fallows, a military-affairs journalist, was more succinct in arguing that "the distinguishing feature of modern American defense has been the pursuit of the magic weapon."[15]

The reformers found much to criticize in U.S. weapon programs. Fallows, for example, termed the F-15 Eagle fighter "a costly unmaintainable system" and the M1 Abrams tank a "cripple."[16] Reformers instead argued that cheaper and simpler weapons were often more effective. Pierre

Sprey, for example, argued that the F-16 Falcon was "clearly more effective" than then F-15 and the M60A1 MBT "clearly superior" to the M1 that replaced it.[17]

In advocating cheaper, simpler weapons, military reformers at times held up Soviet design practices as the model to be emulated. To Edward Luttwak, the Soviets emphasized "raw performance" and a "bold use of advanced materials" while exhibiting an absence of "embellishments." Luttwak praised the Soviet AK-47 assault rifle, which "many practical soldiers insist… is the better weapon by far: it jams less easily, and has a much better feel than the flimsy M-16," a rifle he described as feeling "like a giant toothbrush."[18]

Defense traditionalists countered that advanced technology was a U.S. competitive advantage; forgoing it made no sense. In the words of future Secretary of Defense William J. Perry, "The reform movement's effectiveness is handicapped to the degree that it fails to appreciate the one great advantage that this country has in its competition with the Soviet Union; namely, its technological advantage."[19]

Defense traditionalists believed that it was simplistic to argue that technology and complexity were synonymous. To the contrary, they argued that technology could be harnessed to make weapon systems more reliable and easier to maintain. For example, a *Ticonderoga*-class cruiser possessed twenty times the capability of the ship it replaced and was manned by a crew less than one-third the size. It was both more capable and less expensive to own and operate.[20]

When the claims of the two sides were finally put to the test of combat in the 1991 Gulf War, the results clearly favored the traditionalists. Many of the weapons that performed best under fire, including the M1 Abrams, M2/M3 Bradley, F-15 Eagle, and Advanced Medium-Range Air to Air Missile (AMRAAM), were the same ones the military reformers had long criticized. By contrast, one of the darlings of the military reform movement, the F-16 Falcon, proved to be of limited value because of its short range and inability to drop precision-guided munitions (PGMs). The Iraqis' Soviet hardware, long extolled by defense reformers, proved no match for the U.S. weapons they ridiculed.

Many of the weapons the United States fielded in the late Cold War continue to form the backbone of the U.S. military today. The Army's M1 Abrams Main Battle Tank and M2/M3 Bradley Fighting Vehicle, the Navy's AEGIS cruisers and destroyers and Tomahawk cruise missiles, and the Air Force's F-15 and F-16 fighter aircraft were all developed in response to the Soviet threat but remain first-line weapons in the early twenty-first century.

The Army: From Active Defense to AirLand Battle

The experience of the U.S. Army after Vietnam gives lie to the saying that armies are doomed to fight the last war over again. After the U.S. withdrawal from Vietnam, the Army's leadership turned away from counterinsurgency and focused once again on the confrontation with the Soviet Union in Central Europe. This was in part the result of the painful memory of the war in Southeast Asia. It was also the product of the Army's desire, barely concealed even during the height of the Vietnam War, to plan for high-intensity conventional operations. Soviet military modernization also contributed to this trend. The Soviet Union's deployment of a new generation of weapons and development of revised operational concepts led many leaders to doubt NATO's ability to fight, let alone win, a conventional war in Europe. Several of NATO's military leaders in the 1970s predicted that the alliance would be able to hold out for no longer than ten days before it would be forced to escalate to nuclear use.[21]

In response, a generation of Army officers set about rebuilding the service, both physically and intellectually, to prevent a future replay of Vietnam. They rediscovered strategy, bringing the study of Clausewitz back to the Army War College. They also kindled interest in doctrine and the operational level of war, leading to a renaissance in Army thinking. The result was the development of the doctrine of AirLand Battle, which drove the acquisition of new technology and weaponry, as well as the acquisition of a new generation of weapons, such as the M1 Abrams MBT and M2/M3 Bradley Fighting Vehicle.

Soldiers spend most of their careers studying rather than practicing the profession of arms. This was particularly true during the Cold War, when the superpower confrontation, reinforced by nuclear deterrence, dampened the possibility of major war. It was thus natural that those wars that did break out received close scrutiny. The 1973 Arab-Israeli War was of particular interest to Army officers. A war fought by Israelis who were largely equipped with U.S. arms against Egyptians and Syrians who possessed Soviet weapons, it seemed to offer a close surrogate for a NATO–Warsaw Pact conflict. Moreover, it saw the widespread use of a series of new weapons, such as surface-to-air missiles (SAMs), antitank guided missiles (ATGMs), and antiship cruise missiles (ASCMs). Observers from across globe tried to discern the shape of future wars through the lens of the conflict.

Officers at the U.S. Army's Training and Doctrine Command (TRADOC) studied the war closely. They were struck by the lethality of modern weapons, particularly modern tank guns, ATGMs, and SAMs.[22] As one study concluded, "During the past several decades, the nature of warfare has changed significantly. Great numbers of weapons with increased lethality are found in the armies of both large and small nations. The war in the Middle East in 1973 might well be representative of the nature of future battle. Arabs and Israelis were armed with the latest weapons, and the conflict approached a destructiveness once attributed only to nuclear weapons.... In clashes of massed armor such as the world had not witnessed for 30 years, both sides sustained devastating losses, approaching 50 percent in less than two weeks of combat."[23]

In the aftermath of the 1973 Arab-Israeli War, the Army chief of staff, General Creighton Abrams, dispatched Donn Starry, commander of the Armor Center and School at Ft. Knox, and Brigadier General Bob Baer, the program manager for what became the XM1 MBT, to Israel to study the conflict. Among the lessons that Starry drew were that modern battlefields would be deadly, with greater lethality at greater range and highly lethal air defenses; the result would be enormous equipment losses in a short span of time. Victory would require close cooperation between all combat arms. Perhaps most crucially, they were struck by the importance of seizing and maintaining the initiative.[24]

The 1976 edition of the Army's Field Manual (FM) 100–5, *Operations*, reflected the lessons the Army had drawn from the 1973 War. It contained a stark view of modern warfare, arguing that future conflicts would be characterized by high firepower and attrition.[25] It articulated a new doctrine, dubbed Active Defense, for a future war (more specifically a future war against the Warsaw Pact in Central Europe). The doctrine codified conventional thinking about a NATO–Warsaw Pact war: during the initial phase, NATO forces would be forced onto the defensive, after which they would have to hold out long enough to be reinforced before launching a counterattack.

The appearance of the manual triggered a spirited and often heated debate. Critics decried what they saw as an emphasis on defensive operations and firepower, characterizing Active Defense as "attrition warfare," in contrast to their preferred model of "maneuver warfare." In fact, the political imperative of not surrendering any NATO territory to the Warsaw Pact did much to shape doctrine. More justified was the charge that Active Defense concentrated on the initial battle of a future war and said nothing about follow-on operations.[26]

Dissatisfaction with Active Defense led to the development of a more offensive doctrine, known as "AirLand Battle," which was codified in the 1982 edition of FM 100–5. The manual abandoned Active Defense's focus on direct-fire engagements in favor of strikes deep behind enemy lines. It also emphasized the role of offensive action, maneuver, and surprise.

AirLand Battle was largely the brainchild of Starry, who served as the head of Training and Doctrine Command between 1977 and 1981. The doctrine sought not only to halt an initial Soviet thrust into Central Europe, but also to extend the battle deep into enemy territory. The fact that the Soviets envisioned employing their army in echelons opened opportunities for NATO commanders to use tactical air power and long-range artillery to destroy Soviet armed formations before they made contact with NATO forces.[27]

Starry felt that it was crucial for commanders to see deep into Warsaw Pact territory to locate the follow-on echelon, strike it before the initial assault could break through the NATO defense, and defeat it before it could reach NATO forces.[28] As a result, he envisioned allocating responsibilities to different echelons of command in time rather than distance: brigades would be responsible for attacking all enemy forces within twelve hours of the forward line of troops, divisions within twenty-four hours, and corps within seventy-two hours.[29]

The move from Active Defense to AirLand Battle was a microcosm of a shift in U.S. strategic thinking. The 1980s ushered in a more assertive concept of deterrence, one that envisioned taking the war to Soviet territory from the outset. The Navy's Maritime Strategy, which foresaw operations close to Soviet territory early in a future war, was another manifestation of this shift. Strategists considered other options as well, such as opening new fronts in a NATO–Warsaw Pact war, known as "horizontal escalation."[30]

AirLand Battle drove the development and acquisition of generation of new weapons. To make the doctrine work, the Army required sensors and surveillance systems to warn of attack and identify Soviet forces deep in Warsaw Pact territory, weapons to strike at long range, and a command and control network to link them together. As a result, the Army developed the Joint Surveillance Target Attack Radar System (JSTARS) aircraft to look deep into Eastern Europe and the AH-64 Apache attack helicopter, Pershing II ballistic missile, Multiple-Launch Rocket System, Army Tactical Missile System (ATACMS), and the Copperhead artillery-launched PGM to strike enemy forces. It developed the Tactical Fire Direction (TACFIRE) network to link the systems together.

AirLand Battle had a marked impact on Soviet perceptions of the military balance. The technologies it spawned demonstrated the capacity of the American economy—and American society—to produce weaponry that the Soviets could not match. On a purely military level, Soviet observers saw these weapons as approaching nuclear weapons in effectiveness.

Indeed, some Soviet leaders saw the development of advanced conventional weaponry as presaging a revolution in warfare. As Marshal Nikolai Ogarkov wrote in 1984:

> Rapid changes in the development of conventional means of destruction and the emergence in the developed countries of automated reconnaissance-and-strike complexes, long-range high-accuracy terminally guided combat systems, unmanned flying machines, and qualitatively new electronic control systems make many types of weapons global and make it possible to sharply increase (by at least an order of magnitude) the destructive potential of conventional weapons, bringing them closer, so to speak, to weapons of mass destruction in terms of effectiveness.[31]

American developments demanded a response, one that the Soviet economy was manifestly unable to provide.

Assault Breaker

The need to defeat first- and second-echelon Soviet armored forces during the day or at night and in all types of weather served as an engine of innovation. In 1977, DARPA established Assault Breaker, a program that envisioned using aircraft equipped with a radar that could detect and track vehicular traffic deep in Eastern Europe from high above NATO territory. The aircraft would pass this targeting information to units that would destroy enemy forces with air-launched standoff weapons. The goal of Assault Breaker was to field a system capable of destroying two thousand vehicles operating between twenty and one hundred kilometers behind the front lines in the span of ten hours.[32]

Key to the success of the concept was identifying and tracking vehicles on the ground deep behind enemy lines. In August 1978 the Air Force awarded Grumman and Hughes contracts for a ground moving target indicator (GMTI) radar that became known as Pave Mover. At the same time, the Army was working on a similar system, known as the Standoff Target Acquisition System, or SOTAS. In May 1982, the Director, Defense Research and Engineering (DDR&E) ordered the two programs merged into JSTARS.[33]

The system's airframe was one point of contention. The Army wanted to use the small and slow OV-1D Mohawk, while the Air Force wanted to use the high-flying U-2. Eventually the services settled on Boeing's 707 airframe.[34] Designing the system's software turned out to be a more complex task, requiring almost 600,000 lines of code, nearly three times as many as for the E-3A Sentry Airborne Warning and Control System (AWACS) aircraft. The resulting system could locate and track moving vehicles at a distance of 200 to 250 miles.[35] The aircraft made its first flight in December 1988. Two years later, two test aircraft saw service in the Gulf War. The first production aircraft was delivered in 1996 and became operational in December 1997.[36]

The "Big 5"

Army procurement during the late Cold War focused on five weapon systems, which over time became known as the "Big Five": a state-of-the-art tank, an infantry fighting vehicle (IFV), an advanced attack helicopter, a troop-carrying helicopter, and an air-defense system. Not coincidentally, these programs paralleled the Army's branch structure, with the tank serving the armor community, the IFV serving the infantry and cavalry, the attack helicopter serving aviation, the transport helicopter serving airmobile infantry, and the air-defense system serving air defense artillery.

M1 Abrams

The U.S. Army's TRADOC reflected Army orthodoxy in a 1977 report: "All great armies of the world rest their land combat power upon the tank. The tank, with its cross-country mobility, its protective armor, and its formidable firepower, has been and is likely to remain the single most important weapon for fighting the battle."[37] It is therefore hardly surprising that the centerpiece of Army modernization was a new tank.

The M1 Abrams grew out of two failed attempts to replace the M-60 series of tanks. As discussed in chapter 2, the first was the U.S.-German MBT-70, which became mired in international management problems, conflicting requirements, cost overruns, and technical difficulties. In January 1970 the program collapsed. In its place, Congress authorized an austere version of the tank, which was dubbed the XM803. It was, however, short-lived. The following year opposition to the program began, and Army Chief of Staff William Westmoreland cancelled it in 1971.[38]

Its successor, the XM1, was a departure from previous designs. The Army's requirements called for a 50-ton tank with a crew of four that would be armed with a 105mm gun and would be capable of achieving a top speed of 45 mph.[39] It was the first U.S. tank powered by a gas-turbine engine and featured a fire-suppression system to protect its crew. It also had an advanced digital fire-control system, imaging infrared sensor for the gunner's sight, and a stabilized main gun to allow it to hit targets reliably at longer ranges.[40]

The M1 was equipped with a revolutionary type of armor that was made up of ceramic, steel, and titanium plates laminated between layers of ballistic nylon, a configuration that allowed it to resist a wide variety of antitank rounds. The new armor was a closely held secret; few knew about it or the protection it provided. Its main drawback was its weight: putting it on the XM1 added eight tons to the tank. However, the tank's turbine engine and a new suspension system allowed it to retain its mobility even with the added weight. Fabricating the "special armor," as it was euphemistically known, was another challenge. Its classified composition meant that it had to be manufactured in a secure building, while its exotic constituents required special processes and machines.[41]

The Army released a request for proposals for the XM1 on January 23, 1973. Chrysler and General Motors responded. Extensive discussions and negotiations led to the selection of Chrysler in November 1976.[42] On May 7, 1976, Secretary of Defense Donald Rumsfeld approved the low-rate production of the first batch of 110 XM1 tanks; eventually 2,374 would be produced.[43]

In 1985, the Army began fielding the M1A1, which included a more lethal 120mm smoothbore gun; protection against nuclear, chemical, and biological hazards; and other survivability and habitability improvements.[44] The Army also conducted highly classified research into improving the tank's armored protection. One program involved fabricating armor out of dense depleted uranium. Tests that demonstrated that the armor provided a significant improvement in protection led to the decision to produce an alloy that could be added to the M1's composite armor.[45] Reliably manufacturing the exotic alloy under high-security conditions was a major challenge, however. As a result, early M1A1s were produced without the DU armor. It was introduced in May 1988 in what was called the M1A1 Heavy. Eventually 2,329 M1A1s and 2,140 M1A1 Heavies were produced.[46]

Enhancing the protection of U.S. tanks from Soviet tank rounds and antitank missiles was one part of the equation; enhancing the ability of U.S. tank rounds to penetrate Soviet tank armor was the other. There was considerable concern in the U.S. defense community that Soviet armor developments were outpacing U.S. gun and ammunition technology. In 1982, for example, the CIA assessed that the T-72's armor could defeat any U.S. kinetic-energy tank round or ATGM.[47]

The Gulf War showed such concerns to be overstated. Whereas commentators argued before the war that U.S. forces would have difficulty fighting in the desert, the combination of the flat, open terrain in northern Kuwait and southern Iraq and unimaginative Iraqi tactics allowed

U.S. technology to perform with great effectiveness. U.S. M1A1 tanks possessed a range advantage over the T-72s employed by the Iraqi Republican Guards. Their infrared sights allowed gunners to target Iraqi tanks at night or in blowing sand, situations in which Iraqi tank crews were essentially blind. Even if Iraqi crews had been able to spot American tanks, their cannons lacked the range necessary to strike them. During the Battle of 73 Easting, for example, the Army's 2nd Armored Cavalry Regiment opened fire upon the T-72M1 tanks of Iraq's Tawakalna Republican Guard Division's 18th Mechanized Brigade at 2,400 meters, beyond the 1,800-meter range of the battle sights on the Iraqi Republican Guard T-72s. The ability of U.S. forces to fire while moving prevented the Iraqis from homing in on the muzzle flashes of their tank cannons.[48] Not a single M1 was destroyed or penetrated. Several reported minimal frontal damage despite hits by 125mm smoothbore rounds from Iraqi T-72s.[49] In one stark encounter, a M1A1 "heavy" assigned to the 24th Infantry Division got stuck in the mud and was isolated before coming under attack from 3 Iraqi T-72s. The immobile M1 destroyed all three, including one that was hiding behind a sand berm when it was destroyed. The M1, for its part, was only scratched by a single round from one of the T-72s.

The period following the Gulf War saw additional upgrades to the M1. The M1A2, first rolled out on December 1, 1992, included additional improvements, such as a thermal sensor for the tank commander. More significant was the addition of digital communications, improved navigation capability, and the ability to network tanks together through the Inter-Vehicle Information System.[50] These improvements gave tanks better awareness of the location of both friendly and enemy forces on the battlefield.

M2/M3 Bradley Fighting Vehicle

The late Cold War also saw the modernization of infantry fighting vehicles with the acquisition of the M2/M3 Bradley Fighting Vehicle. The development of fast-moving armored units led to the mechanization of the infantry to allow them to keep up. Infantry units originally rode to the battlefield in trucks, then dismounted to fight. Eventually, armies developed armored personnel carriers for the same purpose. The advent of nuclear weapons called this approach into question; soldiers needed the ability to fight while mounted. However, as mechanized infantry units came more and more to resemble armored forces, infantrymen faced a fundamental question: were those who served in mechanized units infantrymen who rode into combat in armored vehicles, or were they armored forces who fought dismounted?[51]

In November 1967, the Soviets displayed the BMP-1 Mechanized Infantry Combat Vehicle (MICV) for the first time. Armed with a 73mm smoothbore gun, a launch rail for the AT-3 Sagger ATGM, and a coaxial 7.62mm machine gun, the BMP possessed much more firepower than existing APCs. It was also equipped with ports to allow infantrymen to fire their weapons while protected by its armor. In response to the BMP, the U.S. Army launched the XM723 MICV program. In March 1977, the service split the program into the XM2 infantry fighting vehicle and XM3 cavalry fighting vehicle.[52]

The Bradley Fighting Vehicle entered production in 1981 and service in 1983. The M2 became the Army's standard mechanized infantry vehicle, while the M3 replaced the M113 for reconnaissance and security missions in armored cavalry units. The Bradley weighed more than twenty-four tons and was capable of reaching forty mph on the road. Although constructed of aluminum, its laminate side armor provided increased protection over the M113. It was also more heavily armed than its predecessor, with a 25mm machine gun and tube-launched, optically tracked, wire command–link guided (TOW) ATGMs equipped with both optical and imaging infrared sights.

The Bradley was not popular among military reformers. Senators Sam Nunn and Gary Hart questioned the need for the weapon, given its limited protection against tank rounds and antitank missiles. Instead, reformers favored a heavier, tanklike MICV.[53] Although they were unable to kill the program, their concerns did accelerate procurement of the M2A2, with thicker armor and other survivability enhancements.[54]

Desert Storm proved the Bradley to be less vulnerable than its detractors had predicted it would be. On the first day of ground combat, one Bradley from the 3rd Armored Cavalry Regiment was pierced fifteen times by medium and heavy machine-gun rounds and light antitank weapons. Although two scouts aboard were wounded, the vehicle remained operational. The Bradley's reliability exceeded expectations and its weapons proved more capable than predicted.[55]

The net effect of the M1 and M2/M3, combined with the procurement of the AH-64 Apache attack helicopter and ATACMS, was to increase dramatically the lethality of U.S. Army formations: A mechanized division in 1983 had six times the firepower of its World War II predecessor. It could also call upon much more in the way of artillery and air support. Although U.S. heavy divisions superficially resembled the ROAD divisions of the 1960s, they made much greater use of aviation, had ATGMs for use against enemy armor, and had much more lethal artillery, including the Multiple Launch Rocket System (MLRS).[56] These formations, optimized to fight the Warsaw Pact in Central Europe, would see combat against the Iraqi army in 1991 and 2003.

The Navy: Networking and Warfare

The Navy, like the other services, faced the prospect of sharp cuts in the years following the Vietnam War. This manifested itself most vividly in its shipbuilding program. Between 1968 and 1975, the construction of new ships fell by more than two-thirds. As a result, by 1975 the Navy anticipated having to retire 4 percent of the active fleet each year. Budget cuts also led to a reduction in the planned size of the fleet. In 1975, Secretary of Defense James Schlesinger set a goal of a fleet of 575 ships; the following year his successor, Donald Rumsfeld, planned for 600 ships. Both were based upon the requirements of a world war with the Soviet Union. However, the Carter administration questioned the need for such a large fleet, arguing that the Navy's primary use would be in peacekeeping missions and lesser contingencies. Such missions would require a fleet of only 425 to 500 ships.[57]

The Navy also faced more concrete challenges. The growth of the Soviet navy—and particularly of Soviet naval aviation—threatened the ability of the U.S. Navy to operate near the Soviet Union's shores. This challenge forced the Navy to explore networking as well as highly advanced defensive systems such as the AEGIS combat system. Such innovations were fundamentally conservative, meant to preserve the Navy's existing approach to war at sea in the face of an evolving Soviet threat rather than exploring new ways of war.

Networking

During the 1960s, the Soviets deployed a series of increasingly capable ships and submarines. Also of concern were the bombers of Soviet naval aviation, such as the Tu-16 Badger. Paired with long-range air-launched cruise missiles, such as the SS-N-3 Shaddock, these bombers posed a potent threat to the carrier battle groups (CVBGs) that formed the core of the Navy.

The Navy exploited the potential of networking to defend U.S. ships from the Soviet naval threat. In the 1960s, the Navy began developing the Ocean Surveillance Information System (OSIS) as a way to develop a comprehensive system for processing ocean surveillance information. OSIS would collate disparate pieces of information on Soviet naval operations into a

coherent maritime picture. The system tracked Soviet submarines and developed information to help a carrier battle group spot Soviet bombers and
ships early enough to engage them before they could launch their missiles.

The effort benefited from the debut of a new U.S. electronic intelligence
(ELINT) collection system in the autumn of 1976. According to one history, this new source provided a veritable "flood of data" on Soviet naval
activity. At roughly the same time, the U.S. Navy embarked on a program
to correlate the sound characteristics of individual Soviet ships and submarines, a process that became known as hull-to-emitter correlation, or
HULTEC.[58] These advances gave the U.S. Navy a much better understanding of the location and operational patterns of Soviet naval vessels.

To exploit this and other information, the Navy established an OSIS
center at Suitland, Maryland; subsidiary Fleet Ocean Surveillance Information Centers at Norfolk, Virginia; London, England; and Pearl Harbor, Hawaii; and Fleet Ocean Surveillance Information Facilities at Rota,
Spain, and Kamiseya, Japan.[59] At these locations, sailors entered data on
Soviet naval movements into a computer network known as the Navy
Tactical Data System; the data was then correlated and transmitted to
the fleet.[60] The aircraft carrier's Tactical Flag Communication Center
(TFCC) merged this data with real-time information gathered by the
battle group's sensors.[61]

The Navy's networking efforts coincided with the acceleration of the
information revolution. Traditional Navy information systems were cus-
tom built to military specifications. However, the burgeoning commercial
market made more powerful computers available at lower cost. The first
senior officer to exploit the potential of commercial information technology was Rear Admiral Jerry O. Tuttle. In 1981, while serving as a carrier
battle-group commander, Tuttle developed a tactical decision aid using a
package of software applications that were hosted on a commercial desktop computer. The resulting Joint Operational Tactical System, or JOTS,
was, in essence, a TFCC hosted on a commercial computer, providing the
same service without requiring the ship to undergo an expensive overhaul
and installation.[62] As the Navy's Director of Space and Electronic Warfare
from 1989 to 1993, Tuttle was a vigorous advocate of "commercial off-the-
shelf," or COTS, technology. In his view, it was pointless for the Navy to
spend large sums of money developing computers to military specifications when commercial industry could produce better and cheaper machines. He felt that the Navy should use its resources to develop software,
not hardware.

By the early 1990s, virtually all U.S. surface combatants had received
JOTS and its associated terminals. The result was a fleetwide command

and control system known as the Naval Tactical Command System—Afloat (NTCS-A). The adoption of COTS marked a significant change not only in the way the Navy purchased information systems, but also in the flow of information among naval forces. The Navy's traditional approach to networking had been hierarchical and passive: OSIS would develop a picture of the maritime environment and distribute it to the battle group. With JOTS and its successor, the Joint Maritime Command Information System, a distributed network of computers would cooperatively develop the picture. The Navy's networking efforts thus served as a precursor to networking throughout the U.S. armed forces. Indeed, it is hardly a coincidence that the most prominent advocates of networking and "network-centric warfare"—Admiral William Owens and Vice Admiral Arthur Cebrowski—were naval officers.

AEGIS

The combination of long-range bombers and antiship cruise missiles drove other innovations as well. In 1976, the Soviets began deploying the Tu-22M Backfire bomber, an aircraft with twice the range and a much greater payload than the Badger. Naval planners assumed that U.S. carrier battle groups would face one or more regiments of eighteen to twenty-four bombers supported by reconnaissance and electronic warfare aircraft. Ideally, U.S. ships would detect the approach of these bombers in sufficient time to launch fighters and destroy the inbound bombers before they could fire their missiles. The F-14 Tomcat, with its AN/AWG-9 fire-control system and AIM-54 Phoenix air-to-air missile, was designed to intercept and destroy bombers before they could launch their missiles. However, the deployment of the 500-km Kh-22 (AS-4 Kitchen) antiship cruise missile gave Soviet bombers the ability to launch their missiles outside the U.S. air defense envelope.[63]

In 1963, the Navy inaugurated a research program to design an air defense ship to protect CVBGs against the Soviet bomber threat. The result was the AEGIS combat system. Named after the shield of the Greek god Zeus, AEGIS was designed to protect battle groups against antiship missiles that might leak through the outer fighter screen. The heart of the system was an automatic multifunction phased-array radar, the AN/SPY-1. Unlike mechanically steered radars, phased-array radars are steered electronically. As a result, they are able to perform search, track and missile guidance functions simultaneously. The SPY-1, for example, is able to track more than a hundred targets at a time. AEGIS's computer-based command and decision element allowed it to operate against air, sea, and submarine threats.

The AEGIS system was first tested at sea aboard the trial ship USS *Norton Sound* (AVM-1) in 1973. The Navy's first AEGIS ships, the *Ticonderoga*-class cruisers, combined the hull and machinery designs of the *Spruance*-class destroyers with the AEGIS combat system. Additional upgrades were introduced with the USS *Bunker Hill* (CG-52), the first AEGIS ship outfitted with the Vertical Launching System (VLS), which allowed greater firepower. The USS *Princeton* (CG-59) went to sea with the improved AN/SPY-1B radar.

In 1980, the Navy began designing a smaller AEGIS ship with better sea-keeping characteristics, reduced radar and infrared signatures, and an upgraded AEGIS combat system. The first such destroyer was the 8,400-ton USS *Arleigh Burke*, commissioned in 1991. The Navy subsequently purchased more than fifty.

Like other Cold War weapons, AEGIS has outlived the demise of the Soviet Union. Developed to protect carrier battle groups against Soviet bombers and long-range cruise missiles, the AEGIS radar and VLS have become the centerpiece of Navy efforts to defend against ballistic missiles. During the 2003 Iraq War, for example, the AEGIS destroyer *Curtis Wilbur* provided early warning of missile attacks on Kuwait.

Tomahawk

The Navy also improved its striking power through the development of the Tomahawk family of cruise missiles. The Tomahawk, which became a favored method of conducting long-range strikes, had its origins in an antiship cruise missile and nuclear strike system. It offers a case study of how a weapon can be used for missions far different from those that were initially envisioned.

As noted in chapter 1, early cruise missiles were large, vulnerable in flight, and suffered from poor accuracy. However, during the 1960s and 1970s, improvements in engine design, materials, fuel, and guidance transformed the cruise missile into a potent weapon. Of these developments, the most important was the advent of accurate guidance systems using precise inertial navigation systems (INS) and terrain contour matching (TERCOM). First, inertial navigation systems were able to guide missiles much more accurately. Between 1958 and 1970, INS accuracy increased sixfold. Second, U.S. industry developed the ability to use terrain features for missile navigation. Early attempts to use radar terrain matching to guide the Mace cruise missile resulted in failure.[64] However, in 1958 Ling-Temco-Vought (LTV) patented TERCOM, a system that permitted a missile to check its flight profile periodically to determine whether it was on course. It consisted of a radar altimeter and a computer. Stored in the computer were digital altitude profiles of parallel strips of terrain from selected locations along the missile's flight path. As the missile reached an approximate location on the map, the radar altimeter's returns generated a real-time altitude profile, which the computer compared to stored profiles to determine which profile the missile had just flown across.[65]

TERCOM required a large mapping and mission-planning infrastructure. For example, analysts would have needed to gather, evaluate, digitize, and assemble into maps more than one million data points to produce TERCOM maps for a thousand targets.[66] Moreover, those maps relied on imagery from highly classified satellite systems. Tomahawks using TERCOM also required extensive and labor-intensive planning of each mission. The entire flight path of the missile had to be planned in considerable detail and loaded into the missile's computer before it could be fired.[67] Missile flight paths had to be designed to avoid obstacles and air defenses and take advantage of existing TERCOM maps.[68]

To their advocates, cruise missiles had a number of attractive characteristics, including their low cost, ability to be launched by a variety of platforms, and high effectiveness. However, they garnered opposition from each of the U.S. armed services. As Robert J. Art and Stephen E. Ockenden have observed, "The dominant group within each service—the strategic bombers in the Air Force, the carrier admirals in the Navy, and the NATO-conventional arms lobby in the Army—opposed any cruise missile variant that threatened what it conceived to be the service's central mission."[69] The development of cruise missiles would not have occurred as quickly as it did had it not been for the high-level support of the White House, the civilian leadership of the Defense Department, and the State Department. The U.S. political leadership saw cruise missiles as bargaining chips to induce the Soviets to make concessions in arms control negotiations, sweeteners to get Congress and NATO to accept the second Strategic Arms Limitation Talks (SALT II) Treaty, and military options.[70]

What was true of the U.S. armed forces as a whole was true of the Navy in particular. As former Secretary of the Navy John Lehman has written,

> The professional submariners were uncomfortable because [the cruise missile's] primary means of deployment was to be on fleet fast attack submarines. Rightly, the professional focus of the submariner today is on Soviet submarines and not on surface ships, and certainly not on land battles. Therefore, the mission of the Tomahawk was a distraction from their primary responsibilities. Moreover, every Tomahawk aboard left them with one less torpedo to do their primary job, and if it was a nuclear Tomahawk,

they greatly feared that they would be tied to specific firing positions in the event of nuclear alert, frustrating their basic pelagic instincts. The aviators certainly had no love of any system that did not carry a pilot and yet could do some things that carrier aircraft could do. The destroyermen, the surface warfare officers, saw no great benefit from Tomahawk in helping their primary missions of antisubmarine warfare and anti-air warfare.[71]

The Navy's cruise missile program grew out of the need to counter the Soviet surface fleet. Shortly after becoming Chief of Naval Operations, Elmo Zumwalt appointed Admiral Robert Kaufman to chair a panel to explore the possibility of developing a submarine-launched ASCM. The panel recommended the Navy field a tactical antiship cruise missile that would be launched from a new submarine.[72] Secretary of Defense Melvin Laird, for his part, wanted the Navy to develop a nuclear-armed, submarine-launched cruise missile (SLCM) as an insurance policy against the failure of the SALT I negotiations.

The signing of the first Strategic Arms Limitation Talks (SALT I) Treaty in May 1972 left cruise missiles unconstrained. As a result, Laird requested $20 million to launch a cruise missile program as part of a package of amendments to the 1973 defense budget inspired by SALT I. The Defense Department explored a number of options for SLCMs, including vertical and horizontal launch from converted nuclear ballistic missile submarines (SSBNs), launch from SSN torpedo tubes, and vertical launch from a new SSGN. The Navy, which really wanted an ASCM, pressed for another option: a family of cruise missiles with both tactical and strategic applications that would be compatible with existing platforms. This is the view that prevailed. It was not until November 1974, however, that the Navy developed the basic parameters of the new weapon.[73]

The Air Force had its own cruise missile program, which aimed at fielding a nuclear-armed air-launched cruise missile (ALCM). In December 1973, the Defense Department ordered that the services cooperate: it directed the Air Force to share its turbofan engine and high-energy fuel with the Navy, and it told the Navy to share TERCOM with the Air Force.[74]

In March 1976, the Defense Department selected General Dynamics as the manufacturer of the SLCM. The Navy set July 1980 for the initial operational capability of the conventional land attack and antiship variants of the Tomahawk, and January 1981 for the surface-launched conventional variant.[75]

The conventional BGM-109B Tomahawk Antiship Missile (TASM), which was designed to identify and destroy Soviet warships over the horizon, was the missile the Navy wanted. The 460-km missile, equipped with a 1,000-pound warhead, used an inertial navigation system (INS) for navigation and passive and active terminal seekers to home in on targets. It would be launched in the general direction of its target, search it out, identify it, and attack it. However, because it might take the missile half an hour to reach its target, it required in-flight targeting updates. Although the US Navy developed an extensive targeting infrastructure for TASM, over-the-horizon targeting remained the Achilles' heel of the system and it never gained acceptance within the fleet.[76]

Whereas the Navy favored a conventional antiship cruise missile, OSD and the Congress favored the nuclear land-attack version of the Tomahawk, the BGM-109A. Equipped with a W-80 nuclear warhead, the 2,500-km missile would be launched from submarines and ships against shore targets.

As it became apparent how accurately the missile could strike, the Navy decided to field a conventional version of the missile, the BGM-109C. Another version, the BGM-109D, carried bomblets to strike airfields. The conventional Tomahawk Land Attack Missile (TLAM) was the same size, shape, and weight of the nuclear TLAM. However, because it had a heavier warhead, it had a shorter range than the nuclear variant.

Conventional Tomahawks required terminal guidance to correct the inaccuracies that would build up in its INS system over the missile's flight. Early models used a system called Digital Scene Matching Correlation (DSMAC), which compared images of the ground near the target with a digital scene in its memory.[77] In 1993, the Navy introduced the Tomahawk Block III, with an improved engine and warhead. Most significantly, it was equipped with a GPS guidance system. Unlike INS, GPS provided accurate navigational information throughout the missile's flight. More significantly, it made much of the enormous mapping and mission-planning infrastructure that supported Tomahawk obsolete. No longer would TERCOM images be required for missions. This reduced significantly the time needed to plan missions and allowed the missiles to be employed more flexibly. Indeed, the introduction of GPS guidance shows how technological advances can actually make weapon systems simpler rather than more complex.

Both the TLAM and the conventional variant of the ALCM were widely used and highly effective in the wars of the 1990s and beyond. However, the path from their development to their eventual employment was circuitous. First, these were weapons that neither the Navy nor the Air Force wanted when they were in development. Second, the missiles were not used in the roles envisioned when they were developed. The Tomahawk was originally developed as an antiship missile and nuclear land-attack weapon, not a conventional weapon. Finally, their eventual effectiveness was driven by a technology—GPS guidance—that was not originally incorporated in the weapons. The case demonstrates just how hard it can be to predict the utility of weapons even after they have been deployed.

The Air Force: Balancing Quantity and Quality

The Air Force also sought to apply the lessons of Vietnam to its modernization program. Constrained budgets and rising program costs forced the service to adopt a high/low mix: procuring a relatively small number of highly capable (and expensive) aircraft and a larger number of less capable (and inexpensive) ones. In the fighter realm, these were the F-15 Eagle and F-16 Falcon. The Air Force's emphasis on short-range fighters over long-range and ground-attack aircraft reflected both the dominance of the Central Front scenario in defense planning as well as the hegemony of the fighter community within the service.

F-15 Eagle

The F-15 Eagle grew out of the F-X program, which began in 1962. The F-X was originally conceived of as a relatively simple multirole Navy and Air Force aircraft that could be procured in large numbers. Over time, however, the aircraft became larger, more expensive, and more sophisticated as the Air Force added requirements to its design. Navy and Air Force desires also diverged. As neither service wanted a replay of the F-111 debacle, the result was the emergence in 1968 of the F-X as a single-purpose air-superiority fighter for the Air Force. The aircraft's mission coincided with the dominance of fighter pilots within the Air Force. Indeed, the philosophy behind the aircraft's design was "not a pound for air-to-ground."[78]

In September 1968, the Air Force offered eight companies the ability to bid on what had become the F-15 project. The service sought a single-seat, twin-engine air-to-air fighter that emphasized maneuverability, flying qualities, and pilot visibility. It would have a top speed of Mach 2.3 and a third greater range than the F-4. Based on the experience of Vietnam, it would be armed not only with AAMs, but also with a cannon.[79] In December 1969, McDonnell-Douglas was declared the winner of the competition.

The aircraft achieved its first flight in July 1972. In its initial testing, it proved capable of turning tighter than existing fighters and had excellent acceleration, visibility and handling. It also proved highly effective in air-to-air engagements. Its APG-63 radar could detect targets as far away as a hundred nautical miles and had the ability to sort aircraft out from ground clutter, the "look-down, shoot-down" capability. Its major deficiency was its engine, which stalled under certain conditions.[80]

The Air Force deployed its first upgrade of the F-15, the single-seat F-15C and two-seat F-15D, in mid-1979. One major change was the addition of conformal fuel tanks fitted on either side of the fuselage that could carry as much as ten thousand pounds of fuel.[81] The upgrade also had the benefit of highly sensitive intelligence regarding Soviet aerospace programs. From 1977 to 1985, Adolf Tolkachev, a Soviet aerospace engineer, provided the United States designs for the avionics, radar, missiles, and other weapon systems of the MiG-23 Flogger, as well as the missile and radar capabilities of the MiG-25 Foxbat, and he revealed the existence of the Su-27 Flanker and MiG-29 Fulcrum. As a result of this and other information, in December 1979 the Air Force made substantial changes to the electronics package for the F-15, saving the service billions of dollars and up to five years in research and development.[82]

McDonnell-Douglas sought to capitalize on the F-15's potential for air-to-ground missions. It used its own funds to demonstrate a ground-attack variant it dubbed the Strike Eagle. This involved modifying the aircraft's APG-63 radar to give it the ability to provide a high-resolution picture of the ground. It also fitted the aircraft with a Pave Tack pod with a forward-looking infrared (FLIR) sensor for target acquisition and laser designation.[83]

The Air Force pitted the Strike Eagle against a modified F-16, the F-16XL. In February 1984, it declared the F-15E the winner. Four years later, the Air Force took delivery of the first F-15E. It was 13,000 pounds heavier than the F-15C and could carry a 23,500-pound payload. It featured a new cockpit, a new ground-imaging radar (the APG-70), expanded electronic countermeasures, larger wheels and tires, and a digital flight-control system.[84]

F-16 Falcon

Many military reformers were unhappy with the F-15, feeling that the Air Force had added unnecessary complexity, weight, and expense to the aircraft. As a result, they continued to seek the development of an aircraft closer to their ideals, one that was small, simple, reliable, and inexpensive. Such an aircraft would be optimized for close-range air-to-air combat. The Air Force wanted essentially the same thing—a lightweight fighter that could be procured in large numbers to complement the F-15.[85] European NATO members, for their part, were interested in procuring a replacement for the F-104 Starfighter.

In January 1972, the Air Force issued a request for proposals for a 20,000-pound fighter with excellent maneuverability that could be purchased for around $3 million each. The service wanted an aircraft with exceptional pilot visibility that would be armed with a cannon and low-cost AAMs.[86] In response, General Dynamics produced the YF-16, a single-engine fighter distinguished by its bulbous canopy and large engine intake. It was the first aircraft to have a heads-up display (HUD), which allowed a pilot to view targeting data without looking down. The most important innovation, however, was the aircraft's fly-by-wire control system. Standard flight controls transmit commands from the stick and rudder pedals to the aircraft's control services through wires or cables. The F-16's fly-by-wire controls sent the signals electronically.[87]

A flyoff between the YF-16 and Northrop's YF-17, a larger, heavier, two-engine design, led to a decision in January 1975 to acquire the F-16. The YF-17 became the F/A-18 Hornet, which the Navy and Marine Corps later purchased. Six months later, NATO announced the F-16 as the winner of its lightweight fighter competition. Initial plans called for the Air Force to acquire 650 and European NATO members 348. In fact, more than four thousand F-16s have been produced in more than a hundred distinct versions and have seen service in about twenty countries.

Nuclear Modernization: Closing the "Window of Vulnerability"

The late Cold War also witnessed the deployment of new U.S. strategic and intermediate-range nuclear forces in the face of continued Soviet nuclear modernization. However, the acquisition of these systems, particularly the Peacekeeper ICBM and intermediate-range nuclear forces, was rife with controversy. During the 1970s and 1980s, the Soviet Union fielded a new generation of intercontinental ballistic missiles with the payload and accuracy to destroy U.S. ICBMs deployed in hardened silos. Between December 1972 and December 1974, the Soviet Union tested the MR-UR-100 (SS-17 Spanker), a two-stage liquid missile with four multiple independently targeted reentry vehicles (MIRVs).[88] One hundred fifty were eventually deployed. Between April 1972 and October 1975, Moscow tested the UR-100N (SS-19 Stiletto). The initial model (Mod 1) had six 550-kiloton warheads, while the follow on Mod 2 had a single warhead. In

1980, the Soviets began deploying an improved version of the missile, the UR-100NUTTH (SS-19 Stiletto Mod 3), with six MIRVs.[89] Three hundred and sixty were eventually deployed.

Of greatest concern to U.S. planners was the R-36M (SS-18 Satan), a missile with sufficient payload capacity to carry large numbers of heavy warheads. It was tested between February 1973 and October 1975 and was commissioned in December 1975. The SS-18 Mod 1 and Mod 3 were single-warhead versions, while the Mod 2 was equipped with eight MIRVs. Three hundred and eight were eventually deployed. Between 1980 and 1983, the Soviet Union replaced its arsenal of SS-18s with the R-36MUTTH (SS-18 Mod 4), which was equipped with ten MIRVs. At the same time, the Soviets began developing the R-36M2 (SS-18 Mod 5/6), which carried either a single warhead or ten MIRVs.[90]

The Soviet Union also began deploying third-generation SSBNs: quieter boats that were adapted to patrolling beneath the Arctic ice cap and armed with MIRVed SLBMs. Between 1976 and 1982, the Soviet Union launched fourteen Project 667BDR (Delta III) SSBNs armed with sixteen R-29R (SS-N-18) SLBMs. Between 1985 and 1990, the Soviets deployed seven Project 667BDRM (Delta IV) SSBNs with sixteen R-29RM (SS-N-23) missiles, each with up to eight warheads. And between 1981 and 1989, the Soviets launched six Project 941 (Typhoon) SSBNs. The largest SSBN in the world, each boat carried twenty solid-propellant R-39 (SS-N-20) missiles with a range of up to ten thousand kilometers and up to ten MIRVs.[91]

The growth of the Soviet hard-target kill capability spurred efforts to acquire a U.S. hard-target kill capability as well as to ensure the survivability of U.S. nuclear forces through both active and passive defense. Such efforts were, however, proved highly controversial.

Nuclear Modernization

The U.S. effort to hold Soviet ICBMs at risk was an integral part of nuclear force modernization. The Air Force began considering an "Improved Capability Minuteman," the ICBM-X, in 1965. Six years later, SAC issued a requirement for the missile, specifying a large payload and high accuracy; advanced development began in 1973. In August 1978, Secretary of Defense Brown convinced President Jimmy Carter of the need to deploy two hundred ten-warhead MX missiles to counter improvements in the accuracy of Soviet ICBMs.

Designing the missile was one thing; finding a survivable basing mode for it proved to be quite another. Indeed, the quest spanned the terms of three presidents. The combination of arms control, which required veri-fication, and survivability, which required concealment, led to elaborate basing schemes. On September 7, 1979, Carter announced that the MX would use the Multiple Protective Shelter (MPS) system. The scheme envisioned shuttling each of two hundred missiles between twenty-three shelters. Each shelter would be equipped with ports that could be opened periodically to allow Soviet satellites to verify the total number of deployed missiles. The entire complex would occupy 25 million square miles in the remote desert of Nevada and Utah.[92] For understandable reasons, the system generated political opposition among residents of the Southwest.

Two months after taking office, Ronald Reagan appointed a panel to review the basing decision. The panel, however, could not reach a consen-sus, splitting between those who favored basing the missiles deep under-ground and those who advocated deploying them on aircraft that would constantly be airborne. As a compromise, it recommended deploying the first hundred missiles in hardened silos as an interim measure. In October 1981, Reagan cancelled MPS and called for the deployment of a limited number of MX missiles in super-hardened Titan or Minuteman silos in the Midwest.[93]

In May 1982, the idea of deploying ICBMs close together, the "dense pack" scheme, surfaced. The concept was based upon analysis that showed that the blast and debris caused by exploding Soviet re-entry ve-hicles would destroy follow-on weapons, protecting the U.S. missiles. In November, Reagan announced the decision to deploy MX, now dubbed the LGM-118A Peacekeeper, in a dense pack configuration at F. E. Warren Air Force Base, Wyoming.[94] Such a deployment scheme was much more politically palatable than MPS because it would affect only ten to fifteen square miles of territory while also offering a solution to the vulnerability problem. On the other hand, experts disagreed as to whether dense pack would actually work. Some felt that it would only make it easier to destroy the missiles.

As a way out of the vulnerability quandary, some in Congress argued for deploying the so-called Small ICBM or "Midgetman." The concept's supporters argued that it made more sense to deploy large numbers of single-warhead ICBMs than a smaller number of ten-warhead Peace-keepers: single-warhead ICBMs would offer much less lucrative targets and would require the Soviets to expend multiple warheads to guarantee the destruction of a single American warhead.[95] However, because the "Midgetman" lacked a constituency within the Air Force, it never en-tered development.

The Reagan administration also modernized the U.S. SLBM force. Although there was considerable concern over the vulnerability of the U.S. ICBM force, there was less concern over that of U.S. SSBNs. As the U.S. intelligence community concluded in 1976, "We are confident that the Soviets do not now have the capability to determine the location of Western SSBNs at sea with the precision necessary to attack them, or the capability to track them for extended periods."[96]

In November 1981, the Navy began deploying the *Ohio*-class SSBN. The first eight boats were equipped with the Trident C4 missile, but in March 1990, the U.S. Navy deployed the Trident D5, the most accurate SLBM in history. The accuracy was the result of the missile's use of stellar-inertial guidance. This approach to guidance, first used for the never-deployed Skybolt air-launched ballistic missile, was particularly well adapted to mobile missiles such as SLBMs, allowed the missile to determine its initial position. Stellar-inertial guidance was first fielded with the Trident C4 SLBM. The Trident D5 coupled it with the more accurate Mark 6 guidance system and the larger W88 nuclear warhead.[97] The Trident gave the United States the ability to strike hardened targets from the sea for the first time.

Intermediate-Range Nuclear Forces

In the mid-1970s the Soviet Union achieved rough strategic parity with the United States, which made the nuclear balance in Europe all the more important. In the late 1970s, the Soviet Union began replacing older intermediate-range SS-4 and SS-5 missiles with a new intermediate-range missile, the RSD-10 Pioneer (SS-20 Saber), bringing about a perceived qualitative and quantitative change in the European security situation.

The SS-20 was mobile, accurate, and capable of being concealed and rapidly redeployed. It carried three MIRVs, as distinguished from the single warhead carried by its predecessors. The SS-20's five-thousand-kilometer range permitted it to cover targets in Western Europe, North Africa, the Middle East, and, from bases in the eastern Soviet Union, most of Asia, Southeast Asia, and Alaska. Between 1978 and 1986, the Soviet Union deployed 441 of the missiles.[98]

On November 12, 1979, NATO's foreign ministers unanimously adopted a "dual track" strategy to counter Soviet SS-20 deployments. One track called for arms control negotiations between the United States and the Soviet Union to reduce intermediate-range nuclear forces (INF) to the lowest possible level; the other called for deployment in Western Europe, beginning in December 1983, of 464 single-warhead U.S. ground-launched cruise (GLCM) missiles and 108 Pershing II ballistic missiles.

The MGM-31C Pershing II was a substantial modification of the Pershing I IRBM, which had been deployed in Europe since the 1960s. The missile featured new rocket motors with high-energy fuels and lightweight casings to extend its range to 1,800 kilometers. The missile's RV housed a single W85 thermonuclear warhead, with a yield of between five and fifty kilotons. It also had a highly accurate maneuvering re-entry vehicle (MARV) that gave it the ability to destroy, among other targets, Soviet command and control facilities in Eastern Europe and western Russia.

The 1,500-km BGM-109G Ground-Launched Cruise Missile was a derivative of the Tomahawk SLCM. The program was driven more by the need to respond to the Soviet INF deployment than military requirements. Its advocates saw it as survivable, accurate, and cheaper than the Pershing II. It was also a demonstration of U.S. superiority, as the Soviets lacked a comparable system.[99] However, the GLCM ranked low in the Air Force's needs. Moreover, many officers were concerned that manning the weapons would exacerbate the service's personnel shortages.[100]

The Soviets viewed the advent of INF in general, and the Pershing II in particular, as giving the United States new strategic options. The Pershing II's ability to strike Soviet territory, its dramatically improved accuracy, and its reduced yield nuclear warhead led Soviet military planners to conclude that it was designed to kill hardened targets within the Soviet Union. Its eight- to ten-minute flight time rendered obsolete the Soviet air defense system and forced planners to consider major upgrades to strategic defenses.[101]

At a political level, the deployment of the INF missiles in the face of Soviet threats offered concrete evidence of U.S. will and NATO solidarity. The deployment took place in parallel with negotiations to eliminate the weapons. These talks were contentious, precipitating a Soviet walkout on November 23, 1983. Returning to the negotiating table four years late, the Soviets eventually agreed to the elimination of all ground-based ballistic and cruise missiles with ranges between 500 and 5,500 kilometers.[102] The agreement was a milestone, marking the first time the superpowers agreed to eliminate an entire class of weapon.

The Strategic Defense Initiative

The advent of the Strategic Defense Initiative was the most revolutionary development of the late Cold War. It is also the perfect example of an attempt to pit U.S. technology against Soviet industrial might.

The United States continued research into BMD after the demise of Safeguard, though at a reduced level and limited by the ABM Treaty. Of particular interest was the possibility that missile defense could reduce the Soviet threat to U.S. nuclear forces. One concept that the Army explored was to use BMD to enhance the survivability of MX. This involved augmenting the MPS with a mobile BMD system, the Low Altitude Defense System, or LoADS, to protect U.S. missile fields. However, the system would have violated the ABM Treaty's prohibition against mobile BMD systems. It also would have been quite expensive. In October 1980, an Army study estimated that it would cost $8.6 billion over ten years to defend the 4,600-shelter MX deployment.[103]

Safeguard had been based upon a program of nuclear-armed interceptors. However, U.S. research in the years that followed appeared to offer other possibilities. The U.S. Army made advances in the technology to produce a non-nuclear kill vehicle. It also conducted research on directed energy weapons (DEW), such as lasers and particle beams, which would destroy targets by transferring energy to them. The Air Force had its own laser program.[104]

By the beginning of the 1980s, a small number of experts felt that DEW, particularly space-based DEW, offered the prospect of an effective ballistic missile defense system. Senator Malcolm Wallop (R-WY) was an especially prominent proponent. Wallop and other supporters believed that advances in defensive technology meant that a strategic defense system had become feasible. Moreover, space-based DEW, which would destroy Soviet ICBMs and SLBMs early in their flight, held the promise of a system that the Soviets would be unable to overcome merely by adding warheads to their offensive arsenal.[105]

Advocates of strategic defense argued for a fundamental reorientation of U.S. strategy, one that would "move the key competition into a technological arena where we have the advantage. A bold and rapid move into space, if announced and initiated now, would end-run the Soviets in the eyes of the world and move the contest into a new arena where we could exploit the technological advantages we hold."[106] Robert McFarlane, Reagan's national security advisor, supported a strategic defenses program both because it would play to U.S. strength in technology and because it could force the competition with the Soviet Union into an area of U.S. advantage.[107]

Another driver was the belief that the Soviets themselves were pursuing strategic defenses. U.S. experts debated the thrust and extent of the Soviet ballistic-missile-defense effort. Some were concerned that the Soviets might "break out" of the treaty and deploy a nationwide missile defense.

Others worried that the Soviets might "creep out" of the treaty through widespread violations of its restrictions.

Ronald Reagan's announcement of the Strategic Defense Initiative on March 23, 1983, marked a turning point in U.S. defense strategy. In his speech, Reagan called for a shift in thinking from deterrence to defense:

> Let me share with you a vision of the future which offers hope. It is that we embark on a program to counter the awesome Soviet missile threat with measures that are defensive. Let us turn to the very strengths in technology that spawned our great industrial base and that have given us the quality of life we enjoy today.
>
> What if free people could live secure in the knowledge that their security did not rest upon the threat of instant U.S. retaliation to deter a Soviet attack, that we could intercept and destroy strategic ballistic missiles before they reached our own soil or that of our allies?
>
> I know this is a formidable, technical task, one that may not be accomplished before the end of this century. Yet, current technology has attained a level of sophistication where it's reasonable for us to begin this effort. It will take years, probably decades of effort on many fronts. There will be failures and setbacks, just as there will be successes and breakthroughs. And as we proceed, we must remain constant in preserving the nuclear deterrent and maintaining a solid capability for flexible response. But isn't it worth every investment necessary to free the world from the threat of nuclear war? We know it is....
>
> I call upon the scientific community in our country, those who gave us nuclear weapons, to turn their great talents now to the cause of mankind and world peace, to give us the means of rendering these nuclear weapons impotent and obsolete.[108]

For Reagan, at least, strategic defense offered the prospect of absolute security.

SDI proved controversial. Factions within the Air Force and Navy—particularly those involved in nuclear strike—opposed the shift to strategic defense. There was also widespread opposition among the scientific community. Close to seven thousand scientists pledged not to accept SDI money, including the majority of scientists in the physics departments in the top twenty colleges in the United States, as well as fifteen Nobel laureates.[109]

Opponents of SDI, like the opponents of Safeguard a decade earlier, made two somewhat contradictory arguments. First, they argued that building the system was expensive and ultimately infeasible on technical grounds. On the other hand, they argued that such a system, even if only partially effective, would threaten the Soviet Union, causing Moscow to build up its offensive arms and unravel arms control.

SDI proponents envisioned deploying a layered defense against Soviet ballistic missiles. In their view, directed-energy weapons such as ground- and space-based lasers would destroy Soviet ICBMs and SLBMs in their boost phase, before they could deploy their RVs. Space-based directed-energy and kinetic-kill weapons would destroy warheads in their midcourse phase. Finally, high-acceleration ground-based interceptors would destroy warheads as they entered the atmosphere.

Early tests appeared to show that defense against ballistic missiles was indeed feasible. Between February 1983 and June 1984, the U.S. Air Force conducted four "Homing Overlay Experiments" at Vandenberg Air Force Base. The first three tests failed because of mechanical problems, but in the fourth test, officials reported that the interceptor had successfully homed in on, tracked, and destroyed the incoming target missile. The success appeared to show the first proof that defense against ICBMs was within reach.

In fact, the test series was part of a broader effort to deceive the Soviet Union as to U.S. progress in developing strategic defenses. The Defense Department planned to blow the missile up if the interceptor got close enough. However, they did not do so on the first three attempts because the interceptors were so wide of the mark. On the fourth, successful, test, they eliminated the self-destruct mechanism but illuminated the target to increase the chances that the interceptor's infrared sensors would find it.[110]

The deception program was originally designed to block the Soviet Union from gathering accurate information about the U.S. strategic defense program. As it evolved, the program sought to force the Soviets into spending fortunes building their own system and countering that of the United States.[111]

The deception program was discontinued for several reasons. First, it became apparent after the first two misses that it would be hard to portray the SDI system as highly reliable. Second, the risk of Soviet discovery outweighed potential benefits. Third, the deception was increasingly difficult because of the expanding size and complexity of the SDI program. Fourth, the deception was difficult to manage. And fifth and finally, the deception program was a drain on manpower.[112]

The challenge of U.S. advanced technology appears to have had a marked impact on Soviet leaders. In the words of Soviet Ambassador Anatoly Dobrynin, "Our leadership was convinced that the great technical potential of the United States had scored again." Soviet leaders "treated

Reagan's statement as a real threat."[113] The memoirs and recollections of policymakers in Moscow confirm that they took Reagan seriously. An expensive competition in ballistic missile defenses appeared particularly unattractive to Soviet leaders, who were aware of the country's economic difficulties. SDI also highlighted the Soviet Union's lag in computers and microelectronics.[114]

Too much can be made of the role technology played in ending the Cold War. Other dimensions were clearly also important, including the U.S. ideological push against Soviet communism, support to anti-Soviet insurgencies across the globe, and the ossification of the Soviet system itself. But the U.S. technological lead—and its strategic use by Carter and particularly Reagan—cannot easily be dismissed. Soviet military concerns about the widening gap between U.S. and Soviet military technology apparently helped to forge ties between the Soviet political elite and elements of the defense industrial sector on the general need to reorient Soviet foreign policy.[115]

The Cold War guided the development of U.S. military technology for four and a half decades. In the end, however, this technology saw combat not against the Soviets in Central Europe, but against the Iraqis in the desert of Southwest Asia. Decades after they were conceived, developed, and acquired, the weapons of the late Cold War continue to form the backbone of the U.S. armed forces.

Notes

1. Quoted in John B. Hattendorf, *The Evolution of the U.S. Navy's Maritime Strategy, 1977–1986* (Newport, RI: Naval War College Press, 2004), 3.
2. CIA National Foreign Assessment Center, "The Development of Soviet Military Power: Trends Since 1965 and Prospects for the 1980s," SR81–10035X, April 1981, iii–iv.
3. CIA Directorate of Intelligence, "Soviet Military R&D: Resource Implications of Increased Weapon and Space Systems for the 1980s," SOV 83–10064, April 1983, iii.
4. *Soviet Military Power* (Washington, DC: Department of Defense, 1984), 105.
5. Peter Schweizer, *Victory: The Reagan Administration's Secret Strategy That Hastened the Collapse of the Soviet Union* (New York: Atlantic Monthly, 1994), 18.
6. Perry, "Defense Reform," 186–87.
7. National Security Decision Directive 75, "U.S. Relations with the USSR," January 17, 1983, at http://fas.org/irp/offdocs/nsdd-075.htm (accessed October 15, 2004), 2. On the drafting of the document, see Richard Pipes, *Vixi: Memoirs of a Non-Belonger* (New Haven: Yale University Press, 2003), 188–202.
8. CIA Directorate of Intelligence, "The Soviet Defense Industry: Coping with the Military Technological Challenge," SOV 87–10035DX, July 1987, iii.

9. *Soviet Military Power*, 108.

10. Ibid., 109–10.

11. Gus W. Weiss, "The Farewell Dossier," *Studies in Intelligence* (1996) at www.cia. gov/csi/studies/96unclass/farewell.htm (accessed October 14, 2004).

12. Schweizer, *Victory*, 189.

13. General Accounting Office, *Ballistic Missile Defense: Records Indicate Deception Program Did Not Affect 1984 Test Results* (Washington, DC: GAO, July 1994), 3.

14. William S. Lind, "'Quantity Versus Quality' Is *Not* the Issue," *Air University Review* (September–October 1983): 86–88.

15. James Fallows, *National Defense* (New York: Random House, 1981), 35.

16. Ibid.

17. Pierre Sprey, "The Case for Better and Cheaper Weapons," in *Defense Reform Debate*, ed. Asa A. Clark (Baltimore: Johns Hopkins University Press, 1984), 200, 202.

18. Ibid., 107, 98.

19. Perry, "Defense Reform," 192.

20. John F. Lehman, *Command of the Seas* (New York: Scribner's, 1988), 156.

21. Donn A. Starry, "Reflections," in *Camp Colt to Desert Storm: The History of U.S. Armored Forces*, ed. George F. Hofmann and Donn A. Starry (Lexington: University Press of Kentucky, 1999), 546.

22. John L. Romjue, *From Active Defense to AirLand Battle: The Development of Army Doctrine, 1973–1982* (Fort Monroe, VA: Historical Office, U.S. Army Training and Doctrine Command, 1984), 3.

23. Headquarters, United States Army Training and Doctrine Command, "Net Assessment of U.S. and Soviet Tank Crew Training" (Ft. Monroe, VA: U.S. Army Training and Doctrine Command, January 29, 1977), 2–1.

24. Starry, "Reflections," 549.

25. Romjue, *From Active Defense to AirLand Battle*, 7.

26. Richard M. Swain, "AirLand Battle," in Hofmann and Starry, *Camp Colt to Desert Storm: The History of U.S. Armored Forces*, 378; Romjue, *From Active Defense to AirLand Battle*, chapter 2.

27. Romjue, *From Active Defense to AirLand Battle*, 33.

28. Swain, "AirLand Battle," 379.

29. Ibid., 383.

30. See, for example, Steven E. Miller, ed., *Conventional Forces and American Defense Policy* (Princeton, NJ: Princeton University Press, 1986).

31. Quoted in Barry D. Watts, *Long-Range Strike: Imperatives, Urgency and Options* (Washington, DC: Center for Strategic and Budgetary Assessments, 2005), 34.

32. Kenneth P. Werrell, *Chasing the Silver Bullet: U.S. Air Force Weapons Development from Vietnam to Desert Storm* (Washington, DC: Smithsonian Books, 2003), 200.

33. Ibid., 200–202.

34. Ibid., 204.

35. Ibid., 204–5.

36. Richard J. Dunn III, Price T. Bingham, and Charles A. "Bert" Fowler, "Unblinking Eye in the Sky: Moving-target Radar Provides True Picture of the Enemy," *Intelligence, Surveillance, and Reconnaissance Journal* 3, no. 7 (August 2004), 112.

37. Headquarters, United States Army Training and Doctrine Command, "Net Assessment," 2–1, 2–2.

38. Robert J. Sunell, "The Abrams Tank System," in Hofmann and Starry, *Camp Colt to Desert Storm*, 433.

39. Ibid., 435–36.

40. W. Blair Haworth Jr., *The Bradley and How It Got That Way: Technology, Institutions, and the Problem of Mechanized Infantry in the United States Army* (Westport, CT: Greenwood, 1999), 83–84.

41. Ibid., 437, 449.

42. Ibid., 444.

43. Ibid., 454.

44. Ibid., 460.

45. Ibid.

46. Ibid., 463.

47. CIA Directorate of Intelligence, "The Soviet T-72 Tank Performance," SW 82–10067X, August 1982, iv.

48. Robert H. Scales Jr., *Certain Victory: The U.S. Army in the Gulf War* (Washington, DC: Brassey's, 1994), 261.

49. Sunell, "The Abrams Tank System," 432.

50. Ibid., 466.

51. Haworth, *The Bradley and How It Got That Way*, 2.

52. Ibid., 80.

53. Ibid., 81.

54. Ibid., 137.

55. Ibid., 142–43.

56. Combat Studies Institute, *Sixty Years of Reorganizing for Combat: A Historical Trend Analysis* (Fort Leavenworth, KS: U.S. Army Command and General Staff College, 1999), 40, 46.

57. Ibid., 9–10.

58. Christopher Ford and David Rosenberg, *The Admirals' Advantage: U.S. Navy Operational Intelligence in World War II and the Cold War* (Annapolis, MD: U.S. Naval Institute Press, 2005), 61–62.

59. Norman Friedman, *Seapower and Space* (London: Chatham, 2000), 175.

60. Ford and Rosenberg, *Admirals' Advantage*, chapter 4.

61. Friedman, *Seapower and Space*, 188.

62. Ibid., 220.

63. Ibid., 237, 234.

64. Ibid., 135–36.

65. John C. Toomay, "Technical Characteristics," in *Cruise Missiles: Technology, Strategy, Politics*, ed. Richard K. Betts (Washington, DC: Brookings Institution, 1981), 37.

66. Ibid., 39.

67. Michael Russell Rip and James M. Hasik, *The Precision Revolution: GPS and the Future of Aerial Warfare* (Annapolis, MD: Naval Institute Press, 2002), 164, 228.

68. Friedman, *Seapower and Space*, 270.

69. Robert J. Art and Stephen E. Ockenden, "The Domestic Politics of Cruise Missile Development, 1970–1980," in Betts, *Cruise Missiles*, 406.

70. Ibid., 394.

71. Lehman, *Command of the Seas*, 169.

72. Art and Ockenden, "Domestic Politics," 384.

73. Ron Huisken, "The History of Modern Cruise Missile Programs," in Betts, *Cruise Missiles*, 86.
74. Kenneth P. Werrell, *The Evolution of the Cruise Missile* (Maxwell AFB, AL: Air University Press, 1996), 154.
75. Ibid., 155.
76. Friedman, *Seapower and Space*, 211.
77. Ibid., 269.
78. Ibid., 60–65.
79. Ibid., 65–66.
80. Ibid., 68–70.
81. Ibid., 73.
82. Milt Bearden and James Risen, *The Main Enemy* (New York: Random House, 2003), 27.
83. Werrell, *Chasing the Silver Bullet*, 75.
84. Ibid., 75.
85. Ibid., 78.
86. Ibid., 83.
87. Ibid., 86–87.
88. Pavel Podvig, ed., *Russian Strategic Nuclear Forces* (Cambridge, MA: MIT Press, 2001), 213.
89. Ibid., 222.
90. Ibid., 218.
91. Ibid., 243.
92. Donald R. Baucom, *The Origins of SDI: 1944–1983* (Lawrence: University Press of Kansas, 1992), 172–74.
93. Ibid., 176.
94. Ibid., 179.
95. Donald Mackenzie, *Inventing Accuracy: A Historical Sociology of Nuclear Missile Guidance* (Cambridge, MA: MIT Press, 1992), 232.
96. Interagency Intelligence Memorandum, "Soviet Approaches to Defense Against Ballistic Missile Submarines and Prospects for Success," NIO IIM 76–012J, March 1976, 2.
97. Ibid., 243, 246, 275.
98. Ibid., 225–26.
99. Werrell, *Evolution of the Cruise Missile*, 203.
100. Art and Ockenden, "Domestic Politics," 408.
101. Dennis M. Gormley, *Double Zero and Soviet Military Strategy* (London: Jane's, 1988).
102. Office of the Undersecretary of Defense for Acquisition, Technology, and Logistics. INF Treaty, Executive Summary, at www.defenselink.mil/acq/acic/treaties/inf/execsum.htm.
103. Kenneth P. Werrell, *Hitting a Bullet with a Bullet: A History of Ballistic Missile Defense* (Maxwell AFB, AL: Air University Press, 2000), 20.
104. Baucom, *The Origins of SDI*, 103, 108–9.
105. Ibid., 129.
106. Quoted in Baucom, *The Origins of SDI*, 165.
107. Ibid., 182.
108. Ronald Reagan, "Announcement of Strategic Defense Initiative," March 23, 1983, at

www.missilethreat.com/resources/speeches/reagansdi.html (accessed August 16, 2005).

109. Werrell, *Hitting a Bullet with a Bullet*, 22.
110. GAO, *Ballistic Missile Defense*, 3.
111. Tim Weiner, "Lies and Rigged 'Star Wars' Test Fooled the Kremlin, and Congress," *New York Times*, August 18, 1993.
112. GAO, *Ballistic Missile Defense*, 15.
113. Quoted in Jeremi Suri, "Explaining the End of the Cold War: A New Historical Consensus?" *Journal of Cold War Studies* 4, no. 4 (Fall 2002): 65.
114. Ibid., 66.
115. Stephen G. Brooks and William C. Wohlforth, "Power, Globalization, and the End of the Cold War," *International Security* 25, no. 3 (Winter 2000/2001): 165–73.

The Gulf War and the Post–Cold War Era, 1991–2001

THE 1991 GULF WAR differed considerably from both the last major war the U.S. armed forces had fought and the war they anticipated having to fight. It occurred neither in the jungles of Southeast Asia nor on the plains of Central Europe, but in the desert of Southwest Asia. The United States faced neither Viet Cong irregulars nor the Warsaw Pact's armored legions, but the Iraqi army.

The Gulf War was the least constrained conflict the United States had fought since World War II. Although Iraq possessed chemical and (as was later discovered) biological weapons, the United States and its allies waged war with Iraq beyond the shadow of the superpower nuclear arsenals. The Soviet Union, far from being antagonistic (and with less than a year to live), passively supported the U.S.-led coalition. As a result, the United States was able to use its conventional dominance to full effect.

If Vietnam demonstrated the limits of the American way of war, the Gulf War appeared to vindicate it. The conflict served as a showcase not only of technologies that had been in existence for several decades, such as precision-guided munitions (PGMs), but also relatively new ones, such as stealth. Its lopsided outcome led many to argue that a revolution in military affairs (RMA) was in progress. In the years that followed, America's technological edge translated into battlefield effectiveness in punitive strikes against Iraq and interventions in the Balkans. The increasing U.S. reliance on air power and PGMs in these conflicts led many to proclaim a "new American way of war." As the next chapter will show, however, such predictions proved to be premature.

New Ways of War

The wars of the 1990s served as a showcase for precision (applied both to weaponry and navigation) and stealth. These capabilities were the result of decades of Cold War research driven by the need to give the U.S. armed forces a technological edge against the numerically superior Warsaw Pact.

Precision-Guided Munitions

As chapter 3 discussed, laser-guided bombs (LGBs) first saw combat on a large scale during the Vietnam War. The United States expended more than 28,000 LGBs in Southeast Asia between 1968 and 1973, mainly against bridges and transportation chokepoints.[1] The Paveway I in particular proved highly successful: During 1969, Air Force crews delivered 1,601 Paveways, and 61 percent of the 2,000-pound bombs scored direct hits.[2]

The next generation, the Paveway II, featured improved electronics and new materials. By using integrated circuits and electronic miniaturization, designers were able to put more complex systems aboard the bombs to not only improve their accuracy and maneuverability, but also increase the number of bombs that could simultaneously be guided.[3] The bomb came in a number of variants, including the 500-pound GBU-12, 1,000-pound GBU-16, and 2,000-pound GBU-10. In Air Force tests between 1973 and 1976, the weapon achieved accuracies as good as eleven to sixty feet.[4]

The continued development of Warsaw Pact air defenses drove the Air Force to search for weapons with greater standoff range. The Paveway III was designed to be a major advance over the first two generations of LGBs. Whereas early models had to lock on to a target before being released, the Paveway III was able to lock on to the laser designator after it was released from the aircraft, allowing an aircraft to deliver its weapon further from the target. Such a capability came at a price, however: the bomb was much more complex than earlier models. As a result, the program resulted in rising costs and slipping schedules. It was not until 1986—a full decade after the program was started—that the family of weapons became operational.[5]

LGBs proliferated because they employed simple and cheap technology: guidance kits attached to conventional bombs that the U.S. military already had in large numbers. The low cost of the programs also meant that they could avoid much of the scrutiny given to big-ticket programs. It also allowed the U.S. armed forces to conduct numerous tests of the weapons. Still, there was great uncertainty over how well the weapons would work in combat.

Some analysts argued that precision weapons held greater promise. The idea that a non-nuclear weapon with a near-zero Circular Error Probable (CEP) could provide an alternative to nuclear weapons was advanced as early as 1975 by, among others, Albert Wohlstetter.[6] More than a decade later, the Commission on Integrated Long-Term Strategy, which Wohlstetter chaired with Fred Iklé, the undersecretary of defense for policy, reiterated that precision weapons could substitute in some cases for nuclear weapons. It also argued that the widespread adoption of precision weapons could have a dramatic impact on the character and conduct of war. The commission's final report argued, "The much greater precision, range and destructiveness of weapons could extend war across a much wider geographic area, make war much more rapid and intense, and require entirely new modes of operation."[7]

Such futuristic predictions aside, as late as 1990 LGBs remained niche weapons for specialized missions, such as destroying bridges or hardened targets.

Precision Navigation

The advent of precision warfare involved not only the ability to deliver munitions accurately, but also to locate forces on the face of the earth, on the seas, and in the air. As early as the early 1960s, the Defense Department began to study the concept of using radio signals transmitted from satellites for navigation and positioning. The Navy's Transit system became the nation's first satellite-based navigation network when it became operational in 1964. Developed by the Johns Hopkins University's Applied Physics Laboratory, it consisted of a constellation of seven low-altitude, polar-orbiting satellites. The system was originally developed to allow nuclear ballistic missile submarines (SSBNs) and other vessels to determine their location. Although the system was well suited to ships, it was slow, required a long observation time, and provided only two-dimensional location.[8] As a result, it had limited utility for other military applications.

By the late 1960s, each service was working independently on radio navigation systems. In April 1973, the deputy secretary of defense designated the Air Force the lead agency for consolidating the various concepts into a single system known as the Defense Navigation Satellite System and later the NAVSTAR Global Positioning System (GPS).[9]

The GPS constellation allows users to determine their three-dimensional position by measuring the distance from the user to the precise location of GPS satellites as they orbit. Measuring the distance to four GPS satellites, it is possible to establish one's location in three dimensions to within ten to twenty meters. Between 1978 and 1985 the Air Force launched a total of eleven GPS satellites, built by Rockwell International, atop Atlas F boosters. Although one was lost due to a launch failure and others experienced problems in orbit, many continued to operate beyond their three-year design life—in some cases beyond ten years.[10]

The GPS constellation was not cheap. One estimate pegged the lifetime system cost of satellites and user equipment at over $14 billion. Moreover, the program was not without its difficulties. Because GPS was a support system and not a weapon with a clear mission and history of accomplishments, it was difficult for many to understand its value. Moreover, the fact that GPS was a multiservice program meant that its supporters needed to sell it to each of the services.[11]

In 1983, after Soviet warplanes shot down Korean Air Lines Flight 007, President Ronald Reagan announced that GPS signals would be made available for international civil aviation use once the system became operational. Eight years later, the U.S. government announced that the signals would be available free of charge on a continuous, worldwide basis.[12] Initially, commercially available signals were intentionally degraded to give them an accuracy of one hundred meters. In 1996, however, President William J. Clinton signed a presidential directive eliminating the intentional degradation of nonmilitary GPS signals, thus making the more accurate signal available to civilian users.

GPS is in many ways an untraditional weapon system. It is neither a weapon nor a sensor, but a network that provides a capability. It thus epitomizes the trend away from individual weapons to networks. And although the system itself is expensive, the receivers needed to navigate using its signals are not. As a result, GPS has become ubiquitous both for military and nonmilitary users.

Stealth

The development of signature reduction, commonly known as stealth, is the most significant advance in military aviation technology in recent decades. It has greatly increased the ability of manned aircraft to penetrate enemy air defenses. It allows aircraft to fly above the lethal envelope of infrared-guided SAMs and anti-aircraft artillery (AAA) and, under some circumstances, to perform combat missions over hostile airspace while remaining undetected. As the 1988 report of the Commission on Integrated Long-Term Strategy succinctly put it, "Low-observable technology is revolutionary."[13]

Although the existence of stealthy aircraft only emerged in public in the late 1980s, the need to reduce an aircraft's signature to increase its survivability in the face of enemy air defenses is a more traditional concern. The design of both the U-2 and SR-71 reconnaissance aircraft incorporated techniques to reduce their infrared and radar signatures. By the mid-1970s, however, it had become apparent that it would be extremely difficult to penetrate an increasingly sophisticated Soviet air defense network. The need to do so drove the United States to explore aircraft designs optimized to having the smallest radar cross-section possible.

Aircraft achieve stealth through a number of means: by reducing their radar cross-section (RCS), lowering their infrared signature, becoming quieter, and decreasing their visibility. The first aircraft to be designed with these considerations in mind was the F-117A Nighthawk, which grew out of an Air Force program initiated in the early 1970s to explore the application of stealth to aircraft. In the summer of 1974 the Defense Science Board investigated the problems that NATO aircraft would encounter against Warsaw Pact air defenses in a future war. The experience of the Vietnam and 1973 Arab-Israeli wars showed that ground defenses could inflict heavy losses on attacking aircraft.[14] As a result, the Defense Advanced Research Projects Agency (DARPA) initiated conceptual studies of whether it was possible to build a low-observable aircraft. In the summer of 1975 DARPA requested proposals for what became known as the Experimental Survivable Testbed (XST). In November 1975 Lockheed and Northrop were awarded contracts for approximately $1.5 million to design and produce a full-scale model of a low-observable aircraft for RCS testing.[15]

It was Lockheed's Advanced Development Projects Division, the famed "Skunk Works," that produced the design that became the F-117A. The Lockheed aircraft derived its stealth from its faceted structures, which reflected incoming radar waves, and from special paints and coatings that absorbed radar waves. The aircraft's flat surfaces also simplified the complex task of measuring its radar cross-section.[16]

In April 1976 Lockheed was authorized to proceed with the design, construction, and flight-testing of two demonstrator aircraft as part of a highly classified program known as Have Blue.[17] Over the next several

years, the Have Blue aircraft flew against a variety of radars, including actual Soviet equipment, and proved to be virtually undetectable.[18]

Shortly after his appointment in 1977 as undersecretary of defense for
research and engineering, William J. Perry reviewed the most promising
technologies that DARPA was developing. He immediately latched onto
stealth and sought ways to bring it to the field rapidly by shortening markedly acquisition timelines. Although the Air Force embraced stealth, the
Navy was much more cautious.[19]

In November 1978 the Air Force initiated the development of what
would become the F-117 under the code name Senior Trend.[20] The aircraft
was designed to fill the specialized niche of penetrating heavily defended
airspace and striking high-value targets such as command centers, air defense facilities, and airfields. The aircraft's stealth gave it truly novel capabilities. Beneath its skin, however, it was a relatively conventional aircraft,
with cockpit displays and engines from the F/A-18, a fly-by-wire control
system from the F-16, a brake system of the Grumman Gulfstream II, and
landing gear from the F-15.[21]

Because the F-117 was designed to be stealthy, it lacked radar, which
could have allowed adversaries to detect the aircraft. Instead, it identified targets using infrared sensors mated to two laser designators in a system called Infrared Acquisition and Detection System (IRADS). The pilot
would identify a target on the top infrared (IR) system and pass it to a
downward-pointing IR sensor, which would continue to track it for laser
designation and bomb damage assessment.[22]

In the spring of 1977, the Air Force, with Perry's support, decided to
purchase a few stealth fighters before acquiring a stealth bomber.[23] Lockheed signed a contract for the F-117 in November 1978, and the aircraft
made its first flight only thirty-one months later. The original plan called
for producing twenty aircraft for an estimated cost of $33 million each.
The success of the program led to a threefold expansion of the program,
and between August 1982 and July 1990 Lockheed delivered fifty-nine of
the aircraft to the Air Force.[24]

While the F-117 was a "black," or secret, program, speculation about
the plane abounded. The first report claiming that the government was
developing a small stealthy fighter appeared in the summer of 1975. On
August 22, 1980, in the midst of the presidential election season, Secretary
of Defense Harold Brown announced the United States was developing a
new technology, which "alters the military balance significantly." The rev-
elation, which shocked program managers, was made in part to deflect
criticism of the Carter administration's decision to cancel the B-1 Lancer
bomber program.[25] Throughout the early 1980s, speculation about the

plane grew.[26] Drawings that purported to show the design of the aircraft appeared in newspapers and professional journals. One company even released a model of the "F-19 Stealth Fighter"—though the aircraft bore little resemblance to the F-117. Although such speculation was essentially correct, it missed the F-117's faceted design.[27] Such speculation ended when the Pentagon announced the existence of the program in November 1988 and displayed the aircraft in public in April 1990.

The F-117 demonstrated the feasibility of a stealth aircraft. The next step was to apply signature reduction to a bomber. The B-2 Spirit emerged from an effort to find a replacement for the B-52 that could penetrate the Soviet Union's robust air defenses. In September 1980, the Air Force asked Northrop and Lockheed for formal proposals to build such a bomber. The Lockheed proposal, code-named Senior Peg, resembled a larger version of the faceted F-117. The Northrop proposal, code named Senior Ice (and later Senior C.J.), relied primarily on carefully sculpted curves and rounded surfaces and a flying-wing design reminiscent of the Northrop YB-49 of the late 1940s.[28]

Whereas Lockheed had gained experience in stealth through the development of the F-117, Northrop got its expertise through the Battlefield Surveillance Aircraft, Experimental (BSAX) program, which operated under the code name Tacit Blue. In 1978, Northrop received a contract to build an aircraft using curved surfaces to achieve a low RCS. In addition, the aircraft was to be equipped with a low probability of intercept radar and data link. The aircraft flew 135 missions between 1982 and 1985 and demonstrated that a stealthy aircraft could operate safely close to the forward line of the battlefield without being detected by enemy radar. Although two aircraft were built, the design never entered production. However, the aircraft provided valuable engineering data and validated innovative approaches to stealth that were later used in the B-2.[29]

The development of stealth was a part of a strategy for competing with the Soviet Union. One of the main arguments for going ahead with the B-1 Lancer and later the B-2 was to impose on Moscow the tremendous cost of modernizing the Soviet Union's territorial air defense. As the Commission on Integrated Long-Term Strategy noted in 1988, "Stealth operates on a major Soviet vulnerability: the central role assigned to radar-based air defenses in protecting not only the Soviet Union but Warsaw Pact theater forces."[30] Pentagon analysts noted that the Soviets had historically accorded the highest priority to the defense of the Soviet motherland. As a result, Moscow fielded a robust network of early-warning and fire-control radars, air-defense guns, SAMs, and interceptors to defend Soviet territory. The Soviet government also invested considerable sums in passive defenses and civil defense measures.

The ability to penetrate Soviet airspace in the face of such formidable defenses represented an area of considerable advantage for the United States. As Secretary of Defense Caspar Weinberger put it in 1987, "Low observable technologies promise to increase further the competitive advantage of our bomber force, to such a degree as to make obsolete much of the Soviets' air defense infrastructure." In his view, the ability of the United States to penetrate Soviet air space had already forced the Soviets to invest the equivalent of over $120 billion in strategic air defense.[31] The continuing development of stealth would render the Soviet Union vulnerable and force the Soviet leadership to divert funds from offensive arms to defensive arms, thereby imposing costs on the weak Soviet economy and reducing Moscow's ability to threaten the United States.

Part of what made stealth an effective strategy for competing with the Soviets was the U.S. intelligence community's confidence in its understanding of Soviet stealth and counterstealth research. The United States collected information on Soviet stealth research from a variety of sources, including the aerospace engineer Adolf Tolkachev, who spied for the United States between 1977 and 1985. The CIA believed that the Soviets had a good understanding of U.S. stealth programs but were behind the United States. As one 1984 report concluded, "If they have [a stealth] program under way now, it is probably in the very early stages, and deployment probably would not occur until the 1990s because development of new systems requires about a decade."[32] Moreover, the CIA found no evidence of a Soviet counterstealth program.[33] All intelligence indicated that the United States had a clear, and exploitable, lead in stealth.

The B-2 program, like the F-117, was accorded the very highest security classification. Knowledge of it was restricted to a very small circle. The program was based upon very ambitious design specifications, including long range, high payload, and low observability. The B-2 was originally designed to drop nuclear weapons on fixed targets from high altitude. However, in a sign of the Air Force's concern over the viability of stealth, midway through the design process the Air Force added operating at as low as two hundred to three hundred feet in a terrain-following mode to the list of the bomber's requirements. This change set back the first flight of the aircraft by roughly two years and added an excess of $1 billion to its cost.[34] The Air Force also added the more challenging requirement of attacking "strategic relocatable targets" such as the SS-24 and SS-25 intercontinental ballistic missiles (ICBMs), a mission required that the aircraft be fitted with a large radar to identify such targets.

The stealth bomber program was technologically very ambitious. Although a flying wing design was both aerodynamically efficient and potentially stealthy, it posed severe flight-control challenges. Jack Northrop's flying wing design had proven difficult to control when used for the YB-49 several decades earlier. For the B-2, Northrop used computer fly-by-wire controls to make flight control adjustments many times faster than humans could respond. The bomber's design also relied heavily on new materials: the aircraft's central frame was made out of titanium and its wings out of layer upon layer of radar-absorbing graphite tape cemented with epoxy. Portions of the aircraft would be honeycombed for strength and coated with ferrite-based materials that could be "tuned" to diffuse various radar wavelengths.[35] In the course of developing the bomber, Northrop and its subcontractors had to invent some nine hundred new manufacturing processes.[36] All this required great computing power. At the secret Northrop plant in Pico Rivera, California, the company built the largest computing facility west of the Mississippi River.[37] As a result, the program was expensive, totaling $44.7 billion.

In 1987 the Air Force granted Northrop approval to begin procuring 132 of the stealthy bombers for the strategic nuclear attack mission. The first B-2 was publicly displayed on November 22, 1988, only twelve days after the unveiling of the F-117A. It made its first flight on July 17, 1989, and operational testing continued through June 1997. The testing program revealed a number of problems, particularly ones related to the reliability and maintainability of the bomber's low-observable coatings. The aircraft's low-observable materials require close and constant attention, extensive maintenance, and a time-consuming repair process.[38] This increased the amount of time it took to prepare an aircraft for its next mission and reduced the number of sorties it could undertake in a given period. In addition, when deployed the aircraft needed special shelters for maintenance.

All of these considerations bred controversy. On one hand, supporters of the bomber pointed to its ability to elude air defenses as well as the enormous amount of money the Soviet Union would need to expand to counter it. They also noted that although it complemented the other legs of the U.S. nuclear triad, unlike ICBMs and SLBMs, the B-2 could be retargeted. Bomber critics cited the cost of the program, the risk associated with stealth technologies, and the lessened need for a new nuclear bomber with the end of the Cold War.

The end of the Cold War and the disintegration of the Soviet Union led to a diminution of the B-2 program. The original B-2 requirement of 132 had been established early in the Reagan administration. In 1989, there were serious efforts in Congress, led by House Armed Services Committee chairman Les Aspin, to cancel the bomber altogether. Budget limitations compelled the Bush administration to cut the bomber program to seventy-five aircraft less than six months after the fall of the Berlin Wall. Congress balked at even that level, however, and in 1992 the Bush administration capped the program at twenty. Later the Clinton administration allowed the initial flight test vehicle to be upgraded to a bomber, bringing the total inventory to twenty-one.[39] Given the large research, development, and testing and evaluation bill that was rung up developing the aircraft—totaling nearly $25 billion—spreading it over twenty instead of 132 aircraft caused the cost of each aircraft to go up exponentially to $2.1 billion. The aircraft were so few in number that the Air Force named each of them. Like the capital ships of an earlier age, they were named after states.

The bomber's mission also shifted from nuclear to conventional operations. The bomber's unique GPS-Aided Target System (GATS) provided the capability to use the aircraft's radar to target guided weapons as well as to refine target coordinates provided from external sources. Northrop developed the GPS-Aided Munition (GAM) to be used with GATS. GAM was essentially a precursor to JDAM and used the same technology. Although the GATS/GAM combination was successfully demonstrated in 1996 and achieved an initial operational capability the following year, it was never used in combat.

The Air Force was not the only service to explore stealth. The Navy's A-12 Avenger II Advanced Technology Aircraft (ATA) incorporated stealth as well. The aircraft was designed to fly faster and farther than the A-6E Intruder and to carry a large bomb load in an internal bomb bay. At one time, the Navy planned to buy 620 of the aircraft, the Marine Corps 238, and the Air Force 400.

The A-12 proved to be the most troubled of the stealth programs. The extensive use of composites in the aircraft's design caused its weight to grow to over thirty tons, 30 percent over design specifications and very close to the limit of a carrier aircraft. In addition, its complex inverse synthetic aperture radar proved problematic. As a result, costs skyrocketed. By one estimate, the A-12 would have consumed up to 70 percent of the Navy's aircraft budget.[40] On January 7, 1991, after more than $2 billion in contract payments without receiving a single plane. Secretary of Defense Cheney cancelled the program in what was the largest contract termination in Defense Department history. The decision triggered a prolonged legal battle over contract payments.

The Navy also explored stealth at sea. Beginning in the mid-1980s, the Navy, Advanced Research Projects Agency (ARPA), and Lockheed teamed up to examine a variety of new technologies for surface ships, including ship control, automation, sea-keeping, and signature control. The result was the *Sea Shadow*, a 560-ton stealthy surface vessel with a crew of ten capable of traveling at ten knots. Lockheed built the vessel in complete secrecy. Parts from different manufacturers were brought to Redwood City, California, and assembled inside HMB-1, a fully enclosed submersible dock. One concept called for equipping the vessel with SAMs and using them to protect carrier battle groups from Soviet bombers. As it was, the Navy was cool both to the ship and the concept, which did not sit comfortably within the Navy's culture. Within the Navy, the ship was of greatest interest to the SEALs, which were at the time a marginal community. As a result, the *Sea Shadow* remained a test program.[41]

The 1991 Gulf War

The 1991 Gulf War highlighted the effectiveness of precision and stealth in modern warfare. It also marked the combat debut of the generation of military systems that had matured in the late 1970s and throughout the 1980s, including the Army's "Big Five" post-Vietnam weapon systems: the M1 Abrams main battle tank, M-2/M-3 Bradley Fighting Vehicle, AH-64 Apache attack helicopter, UH-60 Blackhawk helicopter, and Patriot surface-to-air missile.[42] The U.S. military pressed other weapons, still in development, into service, including the E-8 Joint Surveillance Target Attack Radar System, or JSTARS. Because the war's outcome was such a lopsided success—at least at the tactical and operational levels—it influenced the course of the U.S. defense debate throughout the decade of the 1990s.

Iraq invaded Kuwait before dawn on August 2, 1990. Within hours, the Iraqi invaders had broken the back of the Kuwaiti army, forcing the royal family to flee to Saudi Arabia. The Kuwaiti air force managed to keep fighting until its base was overrun the next day. The Iraqi government then shamelessly announced that the Kuwaiti royal family had been overthrown in an internal uprising, and that Iraqi troops had entered the country at the request of the new government.

During the next five and a half months, U.S. and coalition forces poured into the theater to deter Iraq from invading Saudi Arabia and set the stage for offensive operations to evict Iraq from Kuwait. By mid-January, the coalition included nearly 1,800 combat aircraft from twelve countries, a large naval force, and approximately 540,000 ground troops from thirty-one countries. They faced some 336,000 Iraq troops organized into forty-two or forty-three divisions.[43]

The U.S. armed forces used the time to upgrade their weaponry. For example, the Army traded out older M1 Abrams tanks for newer M1A1s equipped with a 120mm gun and a chemical defense system. It also upgraded a number of its Bradley Fighting Vehicles to the A2 model, which included a Kevlar liner to improve crew protection. The Defense Department also used the time to deploy the JSTARS, then in development, to the theater. The aircraft, a heavily modified Boeing 707 equipped with a synthetic aperture radar (SAR) optimized to spot movement on the ground, was designed to help ground commanders identify and target enemy forces up to 150 km from the front line at all hours and in all types of weather. The Defense Department dispatched two of the experimental aircraft and eight ground stations to the theater at the request of Lieutenant General Fred Franks, the commander of the Army's VII Corps. The first aircraft flew in theater on January 14, less than seventy-two hours before the beginning of the Gulf War.[44]

Commentators were divided over how effective the U.S. military would be against Iraq. Many analysts predicted that a war would be protracted and costly to the United States. Joshua Epstein of the Brookings Institution used computer modeling to calculate U.S. casualties at between three thousand and sixteen thousand. Former National Security Advisor Zbigniew Brzezinski somewhat less scientifically forecast twenty thousand casualties, while Patrick Buchanan predicted thirty thousand. Senator Edward Kennedy estimated that there would be some three thousand American casualties per week, while former Secretary of the Navy James Webb warned that the U.S. Army would be "bled dry" in three weeks.[45] And there were even more dire warnings. In a report published on the eve of the Gulf War, a group of analysts operating under the auspices of the U.S. Army War College predicted that Iraq's military was "fully capable of keeping pace with the latest innovations in weapons technology. The officer corps understands and is committed to the conduct of combined arms operations to include the integration of chemical weapons. It commands soldiers who, because of their relatively high education level, are able to carry out such operations... We should ask ourselves whether we are prepared for [war with Iraq]—*in our view we are not* ... to perform competently, our forces must be reconfigured, retrained, and re-equipped"[46] (emphasis added).

Speculation over the course of a war between Iraq and the U.S.-led coalition frequently reflected the authors' assumptions about the importance of technology in modern war. As they had throughout the 1980s, the military reformers argued that new technology was expensive and would result in too few military systems to be effective, that advanced systems would be so complex they would be unreliable, and that systems that worked well in the laboratory would not on the chaotic battlefield. Edward

N. Luttwak, for example, portrayed U.S. forces as encumbered by "fanciful tactics, flashy weapons, and promising gadgets" that had not "been tested under combat conditions."[47] Defense traditionalists, by contrast, argued that the U.S. armed forces' emphasis on advanced technology would give the United States an edge on the battlefield.

Operation Desert Storm began on January 17 with an air campaign that was designed to make the most of U.S. technology. In 1991 Iraq possessed one of the most formidable air defense systems in the world, incorporating SAMs, anti-aircraft artillery, radar, and fighters from the Soviet Union, France, China, and others.[48] Indeed, Baghdad was the second most heavily defended city in the world, behind only Moscow.[49] As Air Force planners saw it, stealth and precision would allow U.S. forces to engage in "parallel operations" by attacking a targets deep inside Iraq without rolling back Iraqi air defenses or achieving air superiority. By employing the stealthy F-117, the United States was able to strike at the heart of Iraq before suppressing its air defenses. Indeed, the F-117 was the centerpiece of the strategic air campaign. F-117s attacked with complete surprise and were nearly impervious to Iraqi defenses. During the Gulf War, 36 F-117s attacked the most heavily defended targets in Iraq with GBU-27 2,000 laser-guided bombs. It was the only coalition aircraft that was allowed to strike targets within Baghdad.[50] Although F-117s flew only 2 percent of the total attack sorties in the war, they struck nearly 40 percent of strategic targets such as the leadership and command and control facilities.[51]

The combination of stealth and PGMs gave U.S. forces extremely high effectiveness. A typical non-stealth strike formation in the Gulf War required thirty-eight aircraft, including electronic warfare and defense suppression aircraft, to allow eight to deliver bombs on three aim points. By contrast, only twenty F-117As were able to attack simultaneously thirty-seven targets in the face of more challenging threats. The result was a 1,200 percent increase in target coverage with just over half the aircraft.[52]

As effective as it was, the F-117 was not without its limitations. The aircraft was subsonic and had to operate at night to maximize its stealthiness. Weather was another constraint, affecting nearly one-fifth of its sorties.[53] And even though the aircraft was difficult, if not impossible, for the Iraqis to track, the air commander decided to support its sorties with Air Force EF-111 electronic warfare aircraft.[54]

The United States also used more traditional means to suppress the formidable Iraqi air defense system. Coalition forces made extensive use of antiradiation missiles such as the HARM that home in on SAM radars; it fired more than two thousand of the missiles, including two hundred on the first night alone. In addition, U.S. forces used two types of decoy drones to stimulate the Iraqi radar network and increase its vulnerability to HARM attack. Once Iraqi SAMs were neutralized, coalition aircraft were free to fly above Iraqi anti-aircraft artillery to deliver their ordnance. As a result, coalition air forces suffered casualties less than one-tenth those of the Linebacker II bombing campaign of December 18–29, 1972.[55]

Precision-guided munitions played an important role in the air campaign. One of the dominant images of the war consisted of gun camera footage of LGBs striking Iraqi buildings and bridges with deadly accuracy. The weapons proved dramatically more effective and caused far fewer civilian casualties than in previous wars. It is worth remembering, however, that the effectiveness of PGMs was out of proportion with their numbers: the more than seventeen thousand PGMs expended during the war comprised only 8 percent of the bombs dropped. Indeed, the United States expended three times as many laser-guided bombs in Vietnam than it did in the Gulf War. What was novel was the intensity of PGM use: in six weeks, the coalition dropped more than double the number of LGBs released over North Vietnam in nine months.[56]

Laser-guided bombs were not without their limitations. The most significant was the need to illuminate a target throughout the bomb's flight. Moreover, bombs such as the GBU-24 took between three and eight minutes from target detection to engagement.[57] As a result, clouds and dust could, and did, disrupt operations. On the second night of the war, for example, twenty-two of forty-two LGBs dropped by F-117s missed due to weather; that same night weather prevented F-117s from releasing weapons against another twenty-seven aim points. Moreover, only a small number of aircraft could laser-designate targets, limiting the number of PGMs that could be dropped at any time.[58] Despite the fact that LGBs had been in the U.S. inventory for over two decades, the Air Force had only 118 aircraft with pod-mounted systems for launching and guiding the weapons.[59] Only two types of aircraft—the F-117A and the F-111F—could carry PGMs capable of penetrating hardened targets, such as the GBU-27 or the GBU-24A/B with the I-2000 or BLU-109 warhead.[60] As a result, the campaign against hardened targets such as command bunkers and hardened aircraft shelters was slow and attritional.

The Gulf War vindicated supporters of high-technology aircraft such as the F-111 and F-15 that, owing to their long range, large payload, and PGM capability, were able to make a significant contribution to the air war. By contrast, the aircraft most favored by the military reform movement, the F-16, had less capability because of its short range and limited ability to deliver PGMs. Indeed, the head of the Air Force's official analysis of the war argued that the aircraft's unguided bombs "did little… beyond moving sand around in the Kuwaiti desert."[61]

The war witnessed the innovative use of guided munitions, which were employed not only against fixed strategic targets and hardened aircraft shelters, but also Iraqi armor in revetments. In December 1990, Air Force crews first came up with the idea of "tank plinking," using GBU-12 500-pound LGBs dropped from F-111s to destroy Iraqi tanks protected in earth revetments. The idea was neither part of the F-111F's concept of operations nor something that was planned. Rather, crews discovered that at dusk they were able to spot hot Iraqi tanks on their aircraft's infrared sensors against the cool background of the sand. The first "tank-plinking" sorties occurred on February 5. From then until the end of the war, nearly three-quarters of all F-111F sorties were devoted to enemy ground forces, amounting to 664 sorties in a twenty-three-day period. On the night of February 13–14 alone, for example, forty-six F-111Fs dropped 184 GBU-12s and destroyed 132 Iraqi armored vehicles. Overall, F-111Fs destroyed 920 Iraqi vehicles.[62]

Despite the war's short duration, it nonetheless witnessed technological innovation, including the fielding of the GBU-28, a 5,000-pound laser-guided munition with a penetrating warhead that was used to attack deeply buried hardened Iraqi command centers. The GBU-28 was not even in the early stages of research when Iraq invaded Kuwait. A team from Lockheed and Texas Instruments working at Eglin Air Force Base, Florida, developed it. The team fabricated the weapon beginning February 1, 1991, using surplus Army 8-inch artillery barrels fitted with GBU-27 LGB kits. The barrels were fitted with hardened steel nose cones and tail plugs to protect the weapon's fuse and filled with 650 pounds of molten tritonal explosive. The Air Force approved the program on February 14, and the first units were delivered to the Air Force two days later. Two of the weapons were rushed to the theater and dropped on the final night of the war on a deeply buried command post.[63]

The war also saw the extensive use of cruise missiles, including 282 Tomahawk land attack missiles (TLAMs).[64] TLAM Cs and Ds were the first weapons to strike Baghdad and were the only weapons used for daylight attacks on Baghdad throughout the entire war. Several of them carried spools of a highly classified carbon filament designed to short out the power lines at Iraqi electrical plants.[65] Although the missiles were highly useful, their main drawback was their cost—roughly $1.5 million each.

The war also saw the combat debut of the Conventional Air Launched Cruise Missile, or CALCM, the conventional variant of the nuclear AGM-86B ALCM. The missile was the result of Senior Surprise, a classified Air

Force program initiated in June 1986 to replace the W80 nuclear warhead on a limited number of ALCMs with a 1,000-pound blast fragmentation warhead and the missile's terrain contour-matching guidance system with a GPS system. The ensuing missile, designated AGM-86C, became operational in January 1988.[66]

The Air Force employed the CALCM in the opening night of the war, when seven B-52Hs carrying thirty-nine of the missiles flew nonstop roundtrip from Barksdale Air Force Base, Louisiana, to launch points on the periphery of Iraq. Within a three-and-a-half hour period, the bombers launched thirty-five missiles against eight power generation and transmission and communication sites in Iraq. At the time, the flight was the longest combat sortie in history, covering fourteen thousand miles and thirty-five hours of flight.[67]

The Gulf War air campaign demonstrated U.S. supremacy in the air over Iraq. By the end of the war, coalition aircraft had shot down thirty-three Iraqi fixed-wing aircraft and five helicopters while suffering at most one air-to-air loss. More remarkably, more than 40 percent of kills during the war involved beyond-visual-range engagements, a feat only possible because of the use of Airborne Warning and Control System (AWACS) aircraft, which allowed coalition forces to engage aircraft beyond visual range with little risk of hitting friendly forces. Coalition forces largely dismantled Iraq's formidable KARI integrated air defense system. Coalition air attacks also did substantial damage to the Iraqi army. By the end of the war, the Iraqi army had suffered 76 percent attrition in tanks, 55 percent in armored personnel carriers, and 90 percent in artillery pieces. Although frontline Republican Guard units received less damage than other units, they nonetheless suffered about 50 percent attrition.[68]

The air campaign was not without its flaws, however. During the course of the Gulf War, Iraq launched forty-six modified Scud missiles against Saudi Arabia and forty against Israel in an attempt to inflict U.S. casualties and fracture the coalition.[69] Moreover, coalition aircraft were unable to destroy mobile targets such as missile launchers. There is no evidence that fixed-wing aircraft destroyed a single Scud launcher despite roughly 1,500 strikes against Iraq's missile infrastructure. Although aircrews observed nearly half of Iraq's eighty-eight missile launches, they managed to release their ordnance on only eight occasions and never successfully.[70] Another shortfall was the coalition's failure to destroy Iraq's nuclear, biological, and chemical infrastructure.[71]

Although the use of GPS for navigation received less publicity than precision-guided munitions, it was arguably more important to the overall course of the war. Indeed, it is difficult to disagree with Michael Russell

Rip and James M. Hasik's assessment that GPS was a war-winning technology.[72] The Iraqi and Kuwaiti road network circumscribed the Iraqi military's defensive strategy. Iraqi forces found it hard to navigate offroad.[73] By contrast, GPS allowed coalition forces to navigate, maneuver, and fire with unprecedented accuracy in Kuwait and Iraq's featureless desert despite frequent sandstorms, few paved roads, and few natural landmarks.[74] It also assisted in precision bombing, artillery fire support, and combat search and rescue.

At the time, the GPS constellation had not reached full capability and contained only sixteen satellites, five short of the number needed for full global coverage and eight shy of the complete network. In addition, one of the satellites experienced a failure that threatened the ability of the entire network to support military operations. It remained off-line until operators at Air Force Space Command developed software to bring it back into service. Even with all sixteen satellites operational, there were still seven time windows of up to forty minutes each day during which fewer than the required minimum of four satellites were simultaneously within view of a receiver.[75]

The U.S. armed forces did not fully appreciate the utility of GPS before the war. Before Operation Desert Shield, the Army had only several hundred GPS receivers. On August 20, 1990, however, the Army's Deputy Chief of Operations directed that GPS receivers be sent to the theater as soon as possible. As a result, by mid-January 1991, the Army purchased almost ten thousand commercial units, though less than half that number was sent to the theater.[76] Reliance on commercial GPS receivers meant that the government was forced to turn off the selective availability feature to allow users equipped with commercial receivers to locate their position with the same accuracy as those with military receivers. GPS receivers were not only carried by ground forces, they were also fitted—sometimes with duct tape—to vehicles and helicopters and installed aboard F-16, KC-135, and B-52 aircraft.[77]

GPS was not the only space system to see use in the Gulf War. Indeed, for the first time the United States effectively used all of its space systems in support of military operations. Reconnaissance satellites allowed U.S. forces to identify Iraqi troop concentrations and assess the effectiveness of the air campaign. The Defense Meteorological Support Program's satellites provided weather data needed to plan the air and ground campaigns. And communications satellites linked forces within the theater to their headquarters in Saudi Arabia and the United States.

The Gulf War also witnessed the first wartime battle between ballistic missiles and missile defenses as U.S. Patriot theater ballistic missile defense

(TBMD) batteries attempted to defend Saudi Arabia and Israel from Iraqi missiles. The satellites were able to detect the rocket exhaust plume of Iraqi Scuds within thirty seconds of launch and provide warning of missile attack. Still, the Iraqi missiles' short time of flight—on the order of seven minutes—remained a significant limitation.[78]

One of the primary beneficiaries of Defense Support Program (DSP) warning was the Patriot TBMD system. Patriot was first fielded in 1983 as an anti-aircraft system. In 1984 the Army decided to modify Patriot batteries to allow them to protect point targets such as military bases and airfields against short-range ballistic missiles. It began fielding units equipped with Patriot Advanced Capability (PAC)-1 radar and software modifications in 1988 and missiles with the PAC-2 warhead and fuse modifications in 1990. When Iraq invaded Kuwait there were only three PAC-2 missiles in existence, all experimental.[79] Production of PAC-2 missiles had barely started, and the only operational warhead production line was in Germany.

The missile's manufacturer, Raytheon, undertook a crash program to produce PAC-2 missiles in large numbers. The production line adopted a round-the-clock schedule. As a result, production grew from nine in August to 146 in January.[80]

During the course of the forty-three-day war, U.S. units fired 158 Patriots against fifty-three of the modified Scuds.[81] However, the difficulty of defeating Iraqi missile attacks was manifest in the first attack on Saudi Arabia. Iraq launched five missiles, but by the time they reentered the atmosphere six minutes later at a speed of 4,000 mph, they had broken into fourteen parts, including five warheads. Patriot batteries fired twenty-eight interceptors—two per target—at a cost of $16.8 million.[82] Iraq's modified Scuds also had a tendency to tumble in fight, making intercept more difficult.

On February 25, a Scud slipped through Patriot coverage and slammed into a barracks near Dhahran, killing twenty-eight and wounding ninety-eight in the bloodiest single event of the war. On another occasion, an Iraqi Scud came very close to hitting a pier at the Saudi port of Jubayl that was stacked high with 5,000 tons of 155mm artillery shells. Eight vessels were docked at the pier that day, including two containing materiel for U.S. Marine Corps air wings, several carrying ammunition, the USS *Tarawa* amphibious ship, and a Polish hospital ship.[83]

The years that followed the Gulf War witnessed a lively debate over the effectiveness of the Patriot. An extensive postmortem of the conflict showed that in a number of cases the Patriot's blast-fragmentation warhead hit but did not destroy incoming missiles. As a result, the Army eventually revised downward its original estimates of the missile's effectiveness. Often lost in the debate, however, was the strategic impact of theater missile defense (TMD): even if it did not enjoy high tactical effectiveness it did reassure the Israeli government and population, dampening Israel's incentive to retaliate against Iraq. It also protected U.S. forces in Saudi Arabia.

In the end, the victory over Iraq was surprisingly easy. After thirty-seven days of air combat and a hundred hours of ground combat, Iraq was forced to withdraw from Kuwait, the Kuwaiti government was restored, and Saddam's ability to threaten the region was greatly curtailed. The war's cost was quite light: 146 Americans were killed in the war, and 467 were wounded.[84]

There are many reasons why the U.S.-led coalition defeated Iraq. Some have focused on the disparity between American and Iraqi technology. In their view, Iraqi forces with 1970s and 1980s-vintage Soviet hardware were simply no match for U.S. forces equipped with the latest weaponry. Others have argued that skill was a more important determinant of success. In this view, Iraqi errors created opportunities for U.S. technology to perform effectively. Without these mistakes, the outcome would have been far different, with coalition casualties that "would likely have reached or exceeded prewar expectations" of thousands of deaths, despite the technology gap.[85]

A Revolution in Military Affairs?

The seeming ease with which the U.S.-led coalition defeated Iraq during the Gulf War led many observers in the United States and elsewhere to conclude that the information revolution was bringing about a revolution in military affairs (RMA).[86] The lopsided battles in the deserts of Kuwait and southern Iraq and the seemingly effortless domination of the Iraqi air force indicated to many that warfare had indeed changed. The contrast between prewar expectations of a bloody fight and the wartime reality of Iraqi collapse struck many observers as an indicator of fundamental change.

Some of the more breathless RMA advocates argued that the information revolution marked a complete break with the past. One 1993 report predicted, "The Military Technical Revolution has the potential fundamentally to reshape the nature of warfare. Basic principles of strategy since the time of Machiavelli... may lose their relevance in the face of emerging technologies and doctrines."[87] Others were more cautious. As the authors of the Air Force's official study of the Gulf War concluded, "The ingredients for a transformation of war may well have become visible in the Gulf War, but if a revolution is to occur someone will have to make it."[88]

Although the Gulf War raised the profile of stealth, space systems, and PGMs in the public eye, analysts in both the Soviet Union and the United States had been examining the impact of the information revolution on the conduct of war for over a decade. As chapter 4 discussed, the idea that the emergence of new technology, combined with innovative operational concepts and organizations, would transform the conduct of war first appeared in Soviet military writings in the late 1970s when a group of Soviet officers—led by Marshal Nikolai Ogarkov, the chief of the Soviet General Staff—began arguing that computers, space surveillance, and long-range missiles were changing the character of war.[89] Their main concern was that the United States appeared to be exploiting these technologies much more aggressively than the Soviet Union. They feared that the development of new technology into what the Soviets termed "reconnaissance-strike complexes" would give the United States a significant battlefield edge over the Soviet Union.

Although many ignored or dismissed these writings, analysts in the Pentagon's Office of Net Assessment, led by Andrew W. Marshall, paid serious attention. Marshall saw Soviet concern over what they termed the "military technical revolution" as an opportunity for the United States to influence their behavior. As he later recalled, because the Soviets appeared to be worried that the United States would field "reconnaissance-strike complexes," he felt it might behoove the Defense Department to increase investment in such systems. In early 1991 he commissioned an assessment of how the information revolution might affect warfare. The result was a 1992 report entitled "The Military-Technical Revolution: A Preliminary Assessment," which was circulated within the leadership of the Defense Department, to mainly favorable reviews.[90]

Beginning in 1993, defense experts began talking less about a military-technical revolution (MTR) and more about an RMA. Marshall felt that the former term emphasized technology and while technology makes revolutionary change possible, revolutions take place only when the armed force develop new concepts of operations and create new organizations. In his view, the key task facing the armed forces was not to rush out and purchase new equipment, but to figure out the most appropriate conceptual innovations and organizational changes. He also noted that the information revolution was likely to unfold over the span of decades. He thus felt it best to talk about an "emerging military revolution."[91]

In Marshall's view, there were two plausible ideas about how the information revolution might lead to a revolution in warfare:

> The first is that long-range precision strike weapons coupled to very effective sensors and command and control systems will come to dominate much of warfare. Rather than closing with an opponent, the major operational mode will be destroying him at a distance…. The second idea is the emergence of what might be called information warfare. The information dimension or aspect of warfare may become increasingly central to the outcome of battles and campaigns. Therefore, protecting the effective and continuous operation of one's own information systems, and being able to degrade, destroy, or disrupt the functioning of the opponent's, will become a major focus of operational art.[92]

Marshall warned, however, that the high operational tempo of the U.S. armed forces, the pervasive nature of the information revolution, and the conceptual challenges associated with understanding the information revolution all posed barriers to exploiting the RMA.

The hypothesis that the information revolution was spawning a revolution in military affairs gained high-level support in 1994 with the appointment of William J. Perry as secretary of defense and Admiral William A. Owens as vice chairman of the Joint Chiefs of Staff. As noted earlier, Perry had played an important role in sponsoring many of the technologies associated with the RMA while serving as undersecretary of defense for research and engineering from 1977 to 1981. Owens was a prominent exponent of the view that the U.S. armed forces could realize a major increase in effectiveness by networking together existing weapons, sensors, and command and control systems into a "system of systems." In his view, linking these systems would produce information superiority—or "dominant battlespace knowledge"—and enable a quantum leap in military effectiveness.[93] Both pressed the military to embrace new concepts of operations.

In 1996, the Joint Chiefs of Staff published *Joint Vision 2010*, a document that was supposed to serve as a template for U.S. force modernization. The document argued that technological change could enable a new level of performance across the full range of military operations. It saw information superiority as the key to future military effectiveness, arguing that it would enable four operational concepts: dominant maneuver, precision engagement, full-dimensional protection, and focused logistics.

The congressionally mandated 1997 Quadrennial Defense Review acknowledged the existence of an RMA and committed the department to transforming the U.S. armed forces. As Secretary of Defense William Cohen put it, "The information revolution is creating a Revolution in Military

Affairs that will fundamentally change the way U.S. forces fight. We must exploit these and other technologies to dominate in battle. Our template for seizing on these technologies and ensuring military dominance is *Joint Vision 2010*, the plan set forth by the Chairman of the Joint Chiefs of Staff for military operations of the future."[94]

The National Defense Panel (NDP), mandated by Congress, argued even more strongly in favor of the need to transform U.S. forces. The panel's report argued that an RMA was underway and urged that the Defense Department leadership "undertake a broad transformation of its military and national security structures, operational concepts and equipment, and ... key business processes." The report argued,

> We are on the cusp of a military revolution stimulated by rapid advances in information and information-related technologies. This implies a growing potential to detect, identify, and track far greater numbers of targets over a larger area for a longer time than ever before, and to provide this information much more quickly and effectively than heretofore possible. Those who can exploit these advantages—and thereby dissipate the fog of war—stand to gain significant advantages.... [The Defense Department] should accord the highest priority to executing a transformation of the U.S. military, starting now.[95]

It recommended, among other things, that the department craft a transformation strategy designed to prepare the United States to confront the new and different threats of the twenty-first century. It also argued that the department should place greater emphasis on experimenting with a variety of systems, operational concepts, and force structures.

Although the services embraced transformation rhetorically, they did remarkably little to adapt to the information age. Indeed, in many ways the 1990s represented a lost decade for advocates of transformation. There was a widespread tendency to mouth transformation without making any hard choices. No major acquisition programs were terminated. Instead, their advocates put old wine in new bottles labeled "transformation." Although Strategic Air Command (SAC)—the icon of the nuclear revolution—became U.S. Strategic Command and U.S. Atlantic Command became Joint Forces Command, there was little else in the way of large-scale organizational change. And there were only minor changes in the structure of the armed forces and officer careers.

The armed forces' approach to the information revolution is thus a marked contrast to their exploitation of the nuclear revolution in the 1950s. The nuclear revolution found a very tangible manifestation in nuclear weapons and long-range delivery vehicles. Moreover, it coincided with the beginning of the competition with the Soviet Union, which made exploiting new ways of war an imperative. The revolution also occurred during a period of significant change within the U.S. military, as roles and missions were up for competition. The engine of change for the information revolution, by contrast, was not something as concrete as a single weapon, but rather something as pervasive as information technology. Moreover, the information revolution coincided with the end of the competition with the Soviet Union and questions over the shape of the future security environment. It also took place in an era in which service roles and missions had largely been settled. As a result, the U.S. military experienced much less radical change in the 1990s than the 1950s.

The Rise of Standoff Warfare

The decade following the Gulf War saw the United States use force in the Horn of Africa, the Balkans, and Southwest and Central Asia. Throughout the 1990s, the combination of stealth and precision-guided munitions gave U.S. air forces the ability to strike adversaries from the air with near impunity. In addition, air power seemed uniquely suited to the types of conflicts in which the United States was involved—wars for limited aims fought with partial means for marginal interests. As Eliot A. Cohen put it, "air power is an unusually seductive form of military strength, in part because, like modern courtship, it appears to offer gratification without commitment."[96] Air power coupled with PGMs appeared to offer the ability to coerce Iraq, intervene in the Balkans, and retaliate against terrorist groups while avoiding the difficult decisions associated with a sustained commitment of ground forces.

Iraq

Air power became the favored instrument for dealing with the outcome of the Gulf War. Although coalition troops did a masterful job of ejecting Iraq from Kuwait, the end of the war saw Saddam Hussein still alive, in power, and unrepentant. In the wake of the war, he launched a brutal crackdown against both the Kurds in the north and the Shia Muslims in the south. In April 1991 the United States established a no-fly zone over Iraq above the 39th parallel (Operation Northern Watch) in an effort to shield the Kurds from Baghdad's repression. In August 1992, the United States established a second no-fly zone below the 32nd parallel (Operation

Southern Watch). The result was a protracted game of cat and mouse, as U.S., British, and French forces sought to limit Saddam Hussein's freedom of action while Iraq tried to shoot down coalition aircraft. The operations represented a significant commitment of force. In one year, crews flying out of Incirlik, Turkey, flew more than five thousand sorties during which they attacked some 225 targets. Forces flying *Southern Watch* missions out of Saudi Arabia and Kuwait and off aircraft carriers in the Arabian Gulf sustained an even higher tempo. Between December 1998 and January 2001, Southern Watch aircraft reported coming under fire by Iraqi missiles or aircraft some 670 times.[97] The operation provides yet another illustration of U.S. dominance of the air: by the time the two operations ended in March 2003, the United States had flown sorties over northern Iraq for 4,365 days and over southern Iraq for 3,857 days without suffering a single loss.[98]

Air power also became the preferred method of launching punitive strikes against Iraq. United Nations Security Council Resolution 687, adopted on April 3, 1991, called on Saddam Hussein's regime to destroy its nuclear, chemical, and biological weapons under international supervision; declare all weapons of mass destruction programs; destroy all ballistic missiles with a range greater than 150 kilometers; not commit or support terrorism; cooperate in accounting for missing or dead Kuwaitis; and return property stolen from Kuwait during Iraq's occupation of the emirate. From the start, however, it became clear that Saddam Hussein's regime was half-hearted at best in complying with these conditions. Air strikes presented American decision makers with a relatively cheap, low-risk, and hence attractive option for dealing with Iraqi obstructionism. In early January 1993, after Iraq impeded UN weapons inspectors, American, British, and French aircraft launched strikes against Iraqi air defenses in southern Iraq. In the face of continued Iraqi intransigence, the United States fired forty-five Tomahawks at the Zafaraniyah manufacturing complex outside Baghdad, which had been involved in Iraq's nuclear weapons program. The following day, seventy-five coalition attacks struck additional targets.[99]

In April 1992, the Iraqi government sponsored an assassination attempt on former president George H. W. Bush during his visit to Kuwait. When the plot was uncovered, the Clinton administration ordered a Tomahawk missile strike on the headquarters of the Iraqi Intelligence Service in Baghdad. If the attack was meant to signal American strength and resolve, it failed. Deliberately launched in the middle of the night, the attack destroyed equipment but killed only a janitor.[100]

In September 1996, when Saddam Hussein sent Republican Guard and regular army troops to overrun the Kurdish city of Irbil, the Clinton administration raised the southern no-fly zone to the 33rd parallel and ordered B-52 bombers and Navy warships to launch forty-four cruise missiles against Iraqi air defense sites and command and control facilities in southern Iraq. Although the missiles hit their targets, they did little to aid the beleaguered Kurds in the north. Saddam was able to strike a blow against the Kurds and humiliate the United States without paying a significant price.[101]

In December 1998, when the Iraqi government expelled United Nations inspectors verifying the destruction of Iraq's nuclear, biological, and chemical weapons capability, the United States initiated a three-day air strike known as Operation Desert Fox. Between December 16 and 19, U.S. and British aircraft attacked ninety-seven targets, including command and control facilities, airfields, weapons research facilities, Republican Guard barracks, an oil refinery, and seven of Saddam Hussein's palaces. U.S. combat aircraft flew more than six hundred sorties and U.S. warships launched more than three hundred and thirty cruise missiles, with B-52s launching another ninety. While the ostensible goal of the campaign was to "degrade" Iraq's weapons of mass destruction (WMD) capability and "its ability to threaten its neighbors," it failed to convince the Iraqi government to allow UN inspectors back into the country.[102]

In short, America's dominance in the air, combined with the ability to strike targets with great precision while avoiding to the extent possible harm to innocents, gave the U.S. government a tool to respond to Saddam Hussein's misdeeds. However, these air strikes had little strategic impact upon Iraq. Because they had little effect on Saddam Hussein, they did nothing to influence the underlying source of Iraq's confrontation with the West. It would be left to the George W. Bush administration to deal once and for all with Saddam Hussein's regime.

The Balkans

A second challenge that decision-makers faced in the 1990s was how to respond to the dissolution of Yugoslavia. Air power seemed to provide a means to intervene in Bosnia while avoiding a messy, long-term ground commitment. Beginning in April 1993, U.S. and other NATO aircraft began enforcing a no-fly zone over Bosnia in what came to be known as Operation Deny Flight. In August, NATO began threatening to launch air strikes to punish Bosnian Serbs for laying siege to Sarajevo. In April 1994, NATO aircraft conducted sporadic strikes against Serb targets, but these had only a transitory effect, and in July the Serbs overran the supposed UN "safe haven" of Srebrenica, slaughtering thousands of civilians in the process.[103]

A mortar attack on a Sarajevo marketplace on August 28, 1995, that killed thirty-seven and wounded eighty-five prodded the Clinton administration into calling for air strikes against Bosnian Serbs. The result was Operation Deliberate Force, an air campaign designed to coerce Serbia into negotiating an end to the civil war. The seventeen-day air campaign involved more than 400 aircraft at any one time flying more than 3,500 sorties out of eighteen air bases in five countries and from as many as three aircraft carriers. During the entire campaign, NATO aircraft dropped 1,026 bombs and missiles against forty-eight targets—roughly the same effort as a single day of the Gulf War air campaign.[104]

NATO leaders, concerned by collateral damage, emphasized the use of PGMs over unguided munitions. As a result, 69 percent of the munitions expended during the campaign were precision guided—mostly laser-guided bombs, but also thirteen Tomahawks.[105] As the United States had been in the Gulf War, however, NATO was constrained by the number of aircraft capable of operating at night and in all weather.

The intervention in Bosnia also saw the combat debut of the RQ-1 Predator unmanned aerial vehicle (UAV). Controlled by ground-based operators, these aircraft transmitted electro-optical, infrared, and synthetic aperture radar imagery via satellite to ground stations in the United States or the theater of operations. A unit in Gjader, Albania, operated the Predators that flew over Bosnia. The unit launched the UAV fifteen times, twelve of which were effective, logging over 150 hours of coverage over Bosnia. At one point in the campaign, imagery from the Predator demonstrated that the Serbs were not withdrawing from Sarajevo and led to the decision to continue air strikes.[106]

To some, the air campaign appeared to pay off: In November, all the belligerents met in Dayton, Ohio and agreed to a peace agreement. However, there were other things besides NATO air strikes that caused Serbia to come to the negotiating table. Weeks before Operation Deliberate Force began, the Croatian army launched Operation Storm, a ground offensive aimed at wresting the Krajina region from Serb control. The Croatian offensive dealt the Bosnian Serb army its first major defeat since the war in Bosnia had begun and tipped the military balance against the Bosnian Serbs. It also influenced Serb decision making.

Regardless of the extent to which Croat advances on the ground—and not NATO's strikes from the air—caused the Serb government to come to the table, the outcome of Operation Deliberate Force appeared to vindicate the views of Air Force officers who felt that air power could win wars. The United States would face another test of this proposition in 1999 over Kosovo. Operation Allied Force, NATO's air war over Serbia, came after more than a year of failed attempts to find a negotiated way to stop Serbia's organized repression of Kosovo's ethnic Albanian majority. The air campaign, which stretched from March 24 to June 10, 1999, sought to compel the Serb leadership to quit Kosovo by attacking Serb forces in the province and high-priority targets within Serbia. Over the course of seventy-eight days, NATO aircraft flew just over 38,000 sorties involving 829 aircraft flying from forty-seven locations in Europe and the United States. During the war, NATO aircraft expended over 23,600 munitions against targets in Serbia.[107] The operation ended when Slobodan Milosevic finally acceded to NATO's demands and began withdrawing from Kosovo.

The war began as a demonstration of NATO resolve. Many NATO leaders assumed that Milosevic would back down as soon as he saw that the alliance was serious. The first attack of the war came from cruise missiles fired from four U.S. surface ships and two U.S. and one British attack submarine and six B-52s flying outside Yugoslav airspace.[108] This was followed by strikes by 214 US and 130 Allied aircraft that dropped just over a hundred LGBs.[109]

By the fourth day of the war, it had become clear to NATO leaders that the air offensive had not compelled Milosevic into quitting Kosovo. As a result, the alliance escalated its attacks, striking a broader spectrum of targets in Serbia as well as the Serb army in Kosovo. Reports of mounting Serb atrocities in Kosovo gave the move additional urgency.[110]

There was widespread skepticism as to whether air strikes would be enough to bring Milosevic to heel. Ralph Peters, a retired Army officer and commentator, argued that routing Serb forces from Kosovo would require 100,000 troops and would lead to brutal fighting in villages and cities throughout the province.[111] However, the Clinton administration appeared to take the option of using ground forces off the table.

The fact that Operation Allied Force was waged by a coalition, particularly one that operated by consensus, placed real constraints on the conduct of the war. Any NATO member could veto a strike on any given target. Moreover, throughout the war the alliance conducted operations in such a way as to minimize collateral damage, avoid friendly losses, and preserve Yugoslavia's infrastructure.

The combination of targeting restrictions and Serb tactics prevented the alliance from fully suppressing Serb air defenses. Rather than taking on NATO directly, the Serbs husbanded their high-altitude SAMs and kept their radars off the air. To avoid casualties, NATO pilots were required to remain above 15,000 feet and to fly in strike packages with electronic warfare and defense suppression aircraft. Such tactics greatly reduced the effectiveness of Serb air defenses: Serb SAMs managed to shoot down only two manned aircraft and damage three more.[112] However, unlike the Gulf War, NATO forces were never able to suppress completely Serb air defenses.

On April 21, NATO escalated the conflict, focusing the air campaign on the pillars of the Milosevic regime, including the political apparatus, media, security forces, and economic system. Through an intensified bombing campaign, NATO leaders hoped to bring home the cost of the war to the regime, its supporters, and the general population. U.S. cruise missiles struck Milosevic's political headquarters, as well as that of his wife. Not long afterwards, a B-2 dropped a 4,700-pound GBU-37 bunker buster on the national command center, a multistory facility buried more than a hundred feet underground.[113] In a near replay of the Gulf War, the United States decided to target the Serb electrical system. In the early hours of May 3, F-117s dropped CBU-104(V)2/B cluster munitions filled with spools of carbon graphite threads on five transformer yards in the Yugoslav power grid. The scattered reels of treated wire unwound in the air, draping power lines like tinsel, causing them to short out and cutting off electricity to 70 percent of the country.[114]

Stealth played an important role in Allied Force, as it had during the Gulf War. Just as the F-117 was the only aircraft allowed over Baghdad, only stealthy F-117 and B-2 aircraft were committed over Belgrade for the first fifty-eight days of the war. The war also saw the combat debut of the B-2. Six of the aircraft flew nonstop missions from Whiteman Air Force Base, Missouri, delivering up to sixteen GBU-31 Joint Direct Attack Munition (JDAM) from 40,000 feet against enemy targets. The B-2's forty-five sorties represented less than half a percent of strike sorties but accounted for 11 percent of the bombs dropped on fixed targets. As in the Gulf War, however, stealth aircraft rarely operated alone. On most nights, B-2s received standoff jamming support from Navy and Marine Corps EA-6B Prowlers. Air Force F-16CJ defense suppression aircraft were also in the air to attack any Serb radars that were active in the area.[115]

The war also demonstrated that stealthy aircraft were not invulnerable. On the fourth night of the war a barrage of SA-3 SAMs downed an F-117 Nighthawk northwest of Belgrade. It appears that a lucky combination of low-technology tactics, adaptation to U.S. tactics, and improvisation allowed Serb air defenses to bring down the aircraft.[116] A "lucky shot" or not, the downing of the aircraft pierced the aura of invincibility that had surrounded the aircraft. Moreover, Serbia reportedly recovered debris from the aircraft and gave it to Russia, whose scientists used it to improve the ability of their air defense systems to detect and shoot down stealth aircraft.[117]

The war saw the increased use of PGMs. Some 29 percent of the 23,315 munitions expended in the campaign were precision-guided.[118] In the early weeks of the campaign, more than 90 percent of the ordnance was guided.[119] And, unlike in the Gulf War, nine in ten of the aircraft employed over Serbia were PGM-capable.[120]

PGMs—particularly those relying upon laser-designation or employing optical guidance—were no panacea, however. Weather conditions over Serbia frequently prevented the alliance from launching air strikes. In other cases, the alliance's rules of engagement prevented aircraft from dropping their ordnance.

Allied Force also saw the first combat use of a new generation of PGMs guided by GPS, including the GBU-31 JDAM. JDAM consists of a $21,000 kit that includes a GPS receiver, sensors, and tailfins that is fitted to a standard Mk 84 BLU-109 2,000-pound bomb or Mk 83 1,000-pound bomb.[121] The weapon is guided first by an inertial navigation system but then uses GPS for accuracy updates. The combination gives the weapon accuracy to within ten to fifteen meters.[122] Unlike laser-guided bombs, GPS-guided munitions could be employed through clouds, at night, and through smoke and do not require target designation. B-2 bombers, the only aircraft then capable of delivering JDAM, dropped a total of 652 of the weapons during the air campaign.[123]

The combination of the B-2 and GPS-guided munitions such as the JDAM proved to be a potent one. On a number of occasions, B-2s used their synthetic aperture radars to spot ground targets on approach, reducing the miss distance of the JDAM to less than half the thirteen meters of unaided JDAMs.[124]

Although PGMs were generally extremely accurate, they did occasionally go off course. In other instances, they were aimed at the wrong target. On May 7, three JDAM intended for a Yugoslav arms agency in Belgrade hit the Chinese embassy instead, landing squarely in the part of the embassy that housed intelligence operatives and killing four. CIA analysts who nominated the target misidentified the embassy on maps of Belgrade. The incident was a tactical error with strategic consequences, triggering a diplomatic crisis between Washington and Beijing, disrupting moves to negotiate an end to the war, and prompting a halt to the bombing of targets in Belgrade for the next two weeks.[125]

The war also saw the continued heavy use of Tomahawk. In all, NATO used 218 of the missiles in the conflict. GPS-guided TLAMs struck nearly half of all government and military headquarters, air defense targets, and electrical power grids. Twenty-six Tomahawks, including ten armed with submunition warheads, were used against mobile targets.[126]

The war also saw increased use of UAVs for reconnaissance and surveillance. UAVs—including Army RQ-5A Hunters, Navy RQ-2Q Pioneers, and Air Force RQ-1A Predators—conducted 496 sorties during the conflict to provide commanders a real-time view of the battlefield. RQ-1A Predators identified dispersed Serb army and paramilitary forces. Using SAR, they could identify troop formations through clouds and smoke.[127] Had the war gone on even a few days longer, the Air Force would have begun using UAVs equipped with laser designators to identify targets for attack.

The war over Kosovo also saw the use of information operations. The U.S. government reportedly launched a covert operation to harass and pressure Milosevic's cronies by faxing and calling them.[128] The war also reportedly featured the first operational use of computer network attack. During the war, a U.S. information operations cell reportedly launched attacks against the Yugoslav air defense command and control system.[129]

Why Milosevic chose to withdraw from Kosovo remains a topic of debate. The argument that damage to the Serb army in Kosovo caused Milosevic to fold is the least credible, particularly in light of the fact that NATO air strikes destroyed a relatively small number of Serb armored vehicles.[130] Some argue that the air campaign itself—particularly after it began focusing on Milosevic's inner circle—forced the Serb leader to quit. Others argue that it was the combination of NATO air strikes and the Kosovo Liberation Army's success on the ground that pressured him into giving up. Others have argued that the loss of Russian support for his regime caused him to back down. Still others argue that the prospect of a ground war forced him to concede. While we may never know conclusively, the most persuasive theory appears to be that Milosevic agreed to withdraw from Kosovo not merely because of the damage NATO had inflicted upon Serbia, but rather because he calculated that his situation would deteriorate further if the conflict continued.[131]

The war's outcome appeared to vindicate generations of air-power theorists. As John Keegan admitted shortly thereafter,

> It was less than three weeks ago that the realisation first dawned on me: air power might actually be winning the Balkan war. I turned the thought round for a while and looked at it from several directions, rather as a Creationist Christian might have done on being shown his first dinosaur bone. I

didn't want to change my beliefs but there was too much evidence accumulating to stick to the article of faith. That article of faith, held by all military analysts outside a few beleaguered departments of air power studies in the Service academies, was that air forces could not, alone, win wars.... It now looks as if air power has prevailed in the Balkans and that the time to redefine how victory in war may be won has come.[132]

To the Air Force leadership, the conflict in Kosovo seemed to confirm that the service was on the right path. As the official Air Force report on the war put it:

> The air war over Serbia showed that the Air Force has embraced the RMA—not only in its acquisition strategies for emerging technologies, but in the way it used those technologies during this conflict.... The United States Air Force ... showed that it is a leader in the revolution in military affairs by leveraging new concepts to support future joint and coalition efforts.... The air war over Serbia offered airmen a glimpse of the future, one in which political leaders turned quickly to the choice of aerospace power to secure the Alliance's security interests without resorting to more costly and hazardous alternatives that would have exposed more men and materiel to the ravages of war.[133]

Conversely, the war highlighted the Army's lack of units that were light enough to move quickly yet heavy enough to strike hard. The experience prodded Army Chief of Staff General Eric Shinseki to launch an effort to reconfigure the Army into a more mobile yet lethal force. In October 1999, he announced a goal of transforming the Army into a medium-weight force capable of deploying a five-thousand-strong brigade anywhere in the world within ninety-six hours. As he put it, "We must provide early-entry forces that can operate jointly, without access to fixed forward bases, but we still need the power to slug it out and win decisively."[134] He designated two brigades at Fort Lewis, Washington, as test beds for exploring new concepts and organizations. These units traded in their tracked M1A1 Abrams tanks and M2 Bradley Fighting Vehicles for wheeled LAV III infantry fighting vehicles leased from Canada. They also developed innovative new tactics and organizations.

The experience of the 1990s led a growing chorus of military officers and defense analysts to conclude that such a pattern constituted a "new American way of war." In their formulation, the United States was less apt to use overwhelming force to overthrow enemies, but rather to use incremental force in the pursuit of secondary interests.[135] It seemed to represent a new era in America's use of force. As Air Force Chief of Staff

Ronald R. Fogelman put it in 1996, "America has not only the opportunity but the obligation to transition from a concept of annihilation and attrition warfare that places thousands of young Americans at risk in brute, force-on-force conflicts to a concept that leverages our sophisticated military capabilities to achieve U.S. objectives by applying what I like to refer to as an 'asymmetric force' strategy."[136]

What some saw as a sea change in America's use of force was in fact the byproduct of the strategic environment of the 1990s. As the following chapter will show, the September 11, 2001, terrorist attacks triggered a return to the traditional American way of war, with its use of massive force to overthrow the nation's foes, albeit with means far different from those available to previous generations. In the two years that followed the attacks, the United States and its allies waged wars to overthrow hostile regimes in Afghanistan and Iraq. In doing so, it used the full spectrum of its military capabilities. The main criticism of these operations was not that the United States used too much force for aims that were too expansive, but rather that it should have deployed larger numbers of troops to Iraq in 2003.

Coda: Operation Infinite Reach

The combination of air power and PGMs not only gave the Clinton administration the ability to coerce states, it also gave Washington a way to respond to terrorism. In response to Al Qaeda's 1998 bombing of the U.S. embassies in Kenya and Tanzania, which left twelve Americans and several hundred Africans dead and many more injured, the United States launched a handful of cruise missiles against six terrorist training camps in Afghanistan as well as a pharmaceutical factory in Sudan.

Whereas the attack on Afghanistan was retaliatory, that on Sudan was preemptive, designed to deny terrorists access to chemical weapons. Neither the Sudanese government nor the privately owned Al Shifa pharmaceutical plant was directly implicated in the embassy bombings. Rather, U.S. intelligence agencies suspected that the factory was being used to produce a precursor for the nerve agent VX.[137]

In retrospect, the strike on Afghanistan and Sudan was a bridge between the post–Cold War period and a very new era. In form, it looked a lot like the strikes launched to coerce Iraq and Serbia—limited uses of force to coerce rather than annihilate an adversary. Its target, however, Osama bin Laden's Al Qaeda terrorist network, hinted at the very different war that was soon to come.

Notes

1. Headquarters, U.S. Air Force, Management Information Division, *United States Air Force Statistical Digest: Fiscal Year 1973*, July 31, 1974, table 34, p. 86; *United States Air Force Statistical Digest: Fiscal Year 1974*, April 15, 1975, table 37, p. 73.

2. Kenneth P. Werrell, *Chasing the Silver Bullet: U.S. Air Force Weapons Development from Vietnam to Desert Storm* (Washington, DC: Smithsonian Institution Press, 2003), 149.

3. David R. Mets, *The Quest for a Surgical Strike: The United States Air Force and Laser Guided Bombs* (Eglin AFB, FL: Office of History, Armament Division, Air Force Systems Command, 1987), 98.

4. Werrell, *Chasing the Silver Bullet*, 153.

5. Ibid., 155.

6. Barry D. Watts, *Long-Range Strike: Imperatives, Urgency and Options* (Washington, DC: Center for Strategic and Budgetary Assessments, 2005), 32–33.

7. Fred Ikle and Albert Wohlstetter, *Discriminate Deterrence: Report of The Commission on Integrated Long-Term Strategy* (Washington, DC: U.S. Government Printing Office, 1988), 8.

8. Scott Pace et al., *The Global Positioning System: Assessing National Policies*, MR-614 (Santa Monica, CA: Rand Corporation, 1995), 238.

9. Ibid., 240.

10. Ibid., 242.

11. Ibid., 243, 267.

12. Ibid., 248.

13. Ikle and Wohlstetter, *Discriminate Deterrence*, 49.

14. Werrell, *Chasing the Silver Bullet*, 125.

15. David C. Aronstein and Albert C. Piccirillo, *Have Blue and the F-117A: Evolution of the "Stealth Fighter"* (Reston, VA: American Institute of Aeronautics and Astronautics, 1997), 23, 29.

16. Ben R. Rich and Leo Janos, *Skunk Works* (Boston: Little, Brown, 1994), 21.

17. Aronstein and Piccirillo, *Have Blue and the F-117A*, 33.

18. Werrell, *Chasing the Silver Bullet*, 128.

19. Rick Atkinson, "Stealth: From 18-Inch Model to $70 Billion Muddle," *Washington Post*, October 8, 1989.

20. Aronstein and Piccirillo, *Have Blue and the F-117A*, 2.

21. Werrell, *Chasing the Silver Bullet*, 131–34.

22. Ibid., 132.

23. Rich and Janos, *Skunk Works*, 64.

24. Aronstein and Piccirillo, *Have Blue and the F-117A*, 113.

25. Atkinson, "Stealth."

26. For reports of stealth development in press, see Aronstein and Piccirillo, *Have Blue and the F-117A*, 243–50.

27. Werrell, *Chasing the Silver Bullet*, 135.

28. Atkinson, "Stealth."

29. Werrell, *Chasing the Silver Bullet*, 129–30; Tacit Blue at http://www.wpafb.af.mil/museum/modern_flight/mf37a.htm.

30. Ikle and Wohlstetter, *Discriminate Deterrence*, 49.

31. Caspar W. Weinberger, *Annual Report to the Congress, Fiscal Year 1988* (Washington, DC: U.S. Government Printing Office, 1987).

32. Central Intelligence Agency, Directorate of Intelligence, *Soviet Work on Radar Cross Section Reduction Applicable to a Future Stealth Program*, SW 84–10015, February 1984, iii–iv.

33. CIA Directorate of Intelligence, "U.S. Stealth Programs and Technology: Soviet Exploitation of the Western Press," SW M 88–20036, 1 August 1985, 6.

34. James C. Goodall, *America's Stealth Fighters and Bombers* (Osceola, WI: MBI Publishing, 1992), 69–75.

35. Atkinson, "Stealth."

36. Michael E. Brown, *Flying Blind: The Politics of the U.S. Strategic Bomber Program* (Ithaca, NY: Cornell University Press, 1992), 296.

37. Rick Atkinson, "Unraveling Stealth's 'Black World,'" *Washington Post*, October 9, 1989.

38. Benjamin S. Lambeth, *The Transformation of American Air Power* (Ithaca, NY: Cornell University Press, 2000), 159.

39. Ibid., 158.

40. "A-12 Avenger II," at http://www.fas.org/man/dod-101/sys/ac/a-12.htm (accessed 10 March 2006).

41. Rich and Janos, *Skunk Works*, 274–78.

42. Robert H. Scales, *Certain Victory: The U.S. Army in the Gulf War* (Ft. Leavenworth, KS: U.S. Army Command and General Staff College Press, 1994), 19.

43. Thomas A. Keaney and Eliot A. Cohen, *Gulf War Air Power Survey Summary Report* (Washington, DC: U.S. Government Printing Office, 1993), 7–10.

44. Ibid., 79, 168.

45. Jacob Weisberg, "Gulfballs," *The New Republic*, March 25, 1991.

46. Stephen C. Pelletiere, Douglas V. Johnson II, and Lief R. Rosenberger, *Iraqi Power and U.S. Security in the Middle East* (Carlisle Barracks, PA: U.S. Army War College Strategic Studies Institute, 1990), ix, xi.

47. Edward N. Luttwak, "Blood for Oil: Bush's Growing Dilemma," *The Independent*, August 27, 1990.

48. Michael R. Gordon and Bernard E. Trainor, *The Generals' War* (Boston: Little, Brown, 1995), 87, 102–22, 180–81, 188.

49. Lambeth, *Transformation of American Air Power*, 110.

50. Only F-117s and TLAMs were allowed to attack Baghdad: F-117s at night, and TLAMs during the day. See Keaney and Cohen, *Gulf War Air Power Survey*, 225.

51. Ibid., 224.

52. Lambeth, *Transformation of American Air Power*, 156.

53. Keaney and Cohen, *Gulf War Air Power Survey*, 225.

54. Gordon and Trainor, *The Generals' War*, 118.

55. Ibid., 12, 229–30.

56. Ibid., 226, 241.

57. Michael Russell Rip and James M. Hasik, *The Precision Revolution: GPS and the Future of Aerial Warfare* (Annapolis, MD: Naval Institute Press, 2002), 209.

58. Keaney and Cohen, *Gulf War Air Power Survey*, 15.

59. Rip and Hasik, *Precision Revolution*, 214.

60. Keaney and Cohen, *Gulf War Air Power Survey*, 63.

61. Eliot A. Cohen, "A Bad Rap on High Tech," *Washington Post*, July 16, 1996.

62. Gordon and Trainor, *Generals' War*, 322–23; Lambeth, *Transformation of American Air Power*, 124.

63. Gordon and Trainor, *Generals' War*, 420–21.

64. Keaney and Cohen, *Gulf War Air Power Survey*, 200.

65. Gordon and Trainor, *Generals' War*, 216.

66. Rip and Hasik, *Precision Revolution*, 156.

67. Ibid., 158.

68. Keaney and Cohen, *Gulf War Air Power Survey*, 58, 60, 106.

69. The Iraqis reduced the size of the Scud's 1,000-pound warhead and increased the size of its fuel tank to extend the range of the 190-mile Scud, creating the 400-mile Al Hussein and 550-mile Al Abbas missiles.

70. Watts, *Long-Range Strike*, 52.

71. Keaney and Cohen, *Gulf War Air Power Survey*, 79, 83.

72. Michael Russel Rip and David P. Lusch, "The Precision Revolution: The Navstar Global Positioning System in the Second Gulf War," *Intelligence and National Security* 9, no. 2 (April 1994): 167–241.

73. Gordon and Trainor, *Generals' War*, 353.

74. Pace et al., *Global Positioning System*, 245.

75. Lambeth, *Transformation of American Air Power*, 235–36.

76. Rip and Hasik, *Precision Revolution*, 135–36.

77. Pace et al., *Global Positioning System*, 245.

78. Lambeth, *Transformation of American Air Power*, 237.

79. Scales, *Certain Victory*, 71.

80. Ibid., 72.

81. Stewart M. Powell, "Scud War, Round Two," *Air Force Magazine*, April 1992.

82. Scales, *Certain Victory*, 182–83.

83. Gordon and Trainor, *Generals' War*, 239.

84. Ibid., 457.

85. Stephen Biddle, "Victory Misunderstood: What the Gulf War Tells Us About the Future of Conflict," *International Security* 21, no. 2 (Fall 1996): 139–79.

86. See, for example, William J. Perry, "Desert Storm and Deterrence," *Foreign Affairs* 70, no. 4 (Fall 1991): 66–82; Andrew F. Krepinevich, "Cavalry to Computer," *The National Interest* 37 (Fall 1994): 30–42; Eliot A. Cohen, "A Revolution in Warfare," *Foreign Affairs* 75, no. 2 (March/April 1996): 37–54.

87. Michael J. Mazarr et al., *The Military Technical Revolution: A Structural Framework* (Washington, DC: CSIS, 1993), 28.

88. Keaney and Cohen, *Gulf War Air Power Survey*, 251.

89. See, for example, Stephen J. Blank, "The Soviet Strategic View: Ogarkov on the Revolution in Military Technology," *Strategic Review* (Summer 1984), 3.

90. Andrew F. Krepinevich Jr., *The Military-Technical Revolution: A Preliminary Assessment* (Washington, DC: Center for Strategic and Budgetary Assessments, 2002), i.

91. Andrew W. Marshall, "Some Thoughts on Military Revolutions," memorandum for the record, July 27, 1993.

92. Statement of Andrew W. Marshall, Director, Net Assessment, Office of the Secretary of Defense, before the Senate Armed Services Committee Subcommittee on Acquisition and Technology on May 5, 1995.

93. See, for example, William A. Owens, "The Emerging System of Systems," *U.S. Naval Institute Proceedings* (May 1995): 36–39.

94. William S. Cohen, *Report of the Quadrennial Defense Review* (Washington, DC: Department of Defense, 1997), iv.

95. National Defense Panel, *Transforming Defense: National Security in the 21st Century* (Washington, DC: U.S. Government Printing Office, 1997).

96. Eliot A. Cohen, "The Mystique of Air Power," *Foreign Affairs* 73, no. 1 (January/February 1994): 109.

97. Andrew J. Bacevich, *American Empire* (Cambridge, MA: Harvard University Press, 2002), 152.

98. Wayne Specht, "Iraq Watch Operations Up in the Air," *Stars and Stripes*, March 31, 2003.

99. Kenneth M. Pollack, *The Threatening Storm: The Case for Invading Iraq* (New York: Random House, 2002), 64.

100. Ibid., 67.

101. Ibid., 83.

102. Ibid., 92–94.

103. Karl Mueller, "The Demise of Yugoslavia and the Destruction of Bosnia: Strategic Causes, Effects, and Responses," in *Deliberate Force: A Case Study in Effective Air Campaigning*, ed. Robert C. Owen (Maxwell Air Force Base, AL: Air University Press, 2000); Bacevich, *American Empire*, 163–64.

104. Richard L. Sargent, "Aircraft Used in Deliberate Force" in Owen, *Deliberate Force*, 200, 220.

105. Ibid., 257.

106. Ibid., 227–28.

107. Benjamin S. Lambeth, *NATO's Air War for Kosovo: A Strategic and Operational Assessment* (Santa Monica, CA: Rand Corporation, 2001), 61.

108. Ibid., 21.

109. William M. Arkin, "Operation Allied Force: 'The Most Precise Application of Air Power in History," in *War Over Kosovo: Politics and Strategy in a Global Age*, ed. Andrew J. Bacevich and Eliot A. Cohen (New York: Columbia University Press, 2001), 8.

110. Lambeth, *NATO's Air War for Kosovo*, 25.

111. Ralph Peters, "Invading Kosovo: A Battle Plan," *Newsweek*, May 3, 1999.

112. Lambeth, *NATO's Air War for Kosovo*, 108.

113. Ibid., 187.

114. Ibid., 40–41.

115. Ibid., 92.

116. Ibid., 116–20.

117. David A. Fulghum and Robert Wall, "Russians Admit Testing F-117 Lost in Yugoslavia," *Aviation Week and Space Technology*, October 8, 2001.

118. Lambeth, *NATO's Air War for Kosovo*, 88.

119. Arkin, "Operation Allied Force," 21.

120. Lambeth, *NATO's Air War for Kosovo*, 87.

121. Ross Kerber, "U.S. Bombs Seen Smarter, Cheaper," *Boston Globe*, October 3, 2003.

122. Rip and Hasik, *Precision Revolution*, 236.

123. Lambeth, *NATO's Air War for Kosovo*, 91.

124. Ibid., 90–91.

125. Ibid., 144, 146–47.

126. Michael G. Vickers, "Revolution Deferred: Kosovo and the Transformation of War," in Bacevich and Cohen, *War Over Kosovo*, 194.

127. Ibid., 94–95.

128. Arkin, "Operation Allied Force," 17.

129. Vickers, "Revolution Deferred," 196.

130. Arkin, "Operation Allied Force," 25.

131. Barry R. Posen, "The War for Kosovo: Serbia's Political-Military Strategy," *International Security* 24, no. 4 (Spring 2000): 39–84; Stephen T. Hosmer, *Why Milosevic Decided to Settle When He Did* (Santa Monica, CA: Rand Corporation, 2002).

132. John Keegan, "So the Bomber Got Through to Milosevic After All," *Daily Telegraph*, June 4, 1999.

133. Headquarters United States Air Force, *Initial Report, The Air War Over Serbia: Aerospace Power in Operation Allied Force*, 48–49.

134. Jason Sherman, "Dream Work," *Armed Forces Journal International*, May 2000, 25.

135. See, for example, Eliot A. Cohen, "Kosovo and the New American Way of War," in Bacevich and Cohen, *War Over Kosovo*.

136. Ronald R. Fogelman, "Air Power and the American Way of War," presented at the Air Force Association Air Warfare Symposium, Orlando, FL, February 15, 1996, available at www.au.af.mil/au/awc/awcgate/af/air_power_and_the_american_htm (accessed March 10, 2006).

137. Some have questioned whether the factory was in fact involved with nerve gas production. See James Risen and David Johnston, "Experts Find No Arms Chemicals at Bombed Sudan Plant," *New York Times*, February 9, 1999. See also Daniel Benjamin and Steven Simon, *The Age of Sacred Terror* (New York: Random House, 2002).

The Global War on Terrorism, 2001–2005

THE SEPTEMBER 11, 2001, terrorist attacks and the response to them marked the end of the post–Cold War period and the beginning of a new era. Al Qaeda's use of four hijacked airliners as manned missiles in attacks on the World Trade Center and Pentagon not only killed more than three thousand innocents in New York City, Washington, and Shanksville, Pennsylvania, but also led to the wholesale reorientation of U.S. national security policy. Gone were limited strikes in response to ambiguous threats. Instead, the United States embarked on a campaign to eliminate Al Qaeda and the Taliban regime that harbored them.

Speaking before a joint session of Congress a week after the attacks, President George W. Bush emphasized discontinuity with the past:

> Americans have known wars—but for the past 136 years, they have been wars on foreign soil, except for one Sunday in 1941. Americans have known the casualties of war—but not at the center of a great city on a peaceful morning. Americans have known surprise attacks—but never before on thousands of civilians. All of this was brought upon us in a single day—and night fell on a different world, a world where freedom itself is under attack.[1]

The president sought to sketch the broad outlines of the war to come:

This war will not be like the war against Iraq a decade ago, with a decisive liberation of territory and a swift conclusion. It will not look like the air war above Kosovo two years ago, where no ground troops were used and not a single American was lost in combat. Our response involves far more than instant retaliation and isolated strikes. Americans should not expect one battle, but a lengthy campaign, unlike any other we have ever seen. It may include dramatic strikes, visible on TV, and covert operations, secret even in success.[2]

The September 11 terrorist attack marked another shift, this one involving the American way of war. During the Cold War, the U.S.-Soviet nuclear standoff constrained both the types of force that could be used as well as the aims for which it could be used. During the 1990s, the United States repeatedly resorted to limited strikes—often delivered from the air—to coerce adversaries. As chapter 5 notes, some analysts went so far as to argue that this portended a "new American way of war." The days and weeks that followed the September 11 attacks showed such predictions to have been premature. Indeed, what occurred after the attacks was in many respects a return to the traditional American way of war, with its use of massive force to overthrow the nation's foes, albeit with means far different than those available to previous generations.

This chapter explores the first two campaigns of the long war with Islamic terrorists. It begins with Operation Enduring Freedom in Afghanistan, a campaign in which the United States and its allies used teams of special operations forces (SOF) and Central Intelligence Agency (CIA) paramilitary operatives, working with local forces and backed by air power, to overthrow Afghanistan's repressive Taliban regime and deny Al Qaeda a sanctuary. It then explores the 2003 Iraq War, Operation Iraqi Freedom. Its major combat phase demonstrated just how far the U.S. military had come since the 1991 Gulf War. The insurgency that grew out of the outcome of the war, however, demonstrates both the utility and limits of military technology in battling irregular adversaries.

Afghanistan

The first campaign of the war against Islamic extremists was played out in Afghanistan, which Al Qaeda had for years used as a safe haven. In the wake of the September 11, 2001, terrorist attacks, President George W. Bush asked the National Security Council for plans to overthrow the Taliban and evict Al Qaeda from Afghanistan. The Joint Chiefs of Staff produced a range of options, from cruise missile strikes to the deployment of several Army divisions over months. The CIA produced a much different plan, one that called for a covert war against the Taliban that would feature CIA paramilitary forces and anti-Taliban guerrillas.[3] The latter became the basis of U.S. operations.

On September 25, Secretary of Defense Donald Rumsfeld announced the beginning of Operation Enduring Freedom. The following day, the first CIA team, code-named Jawbreaker, entered Afghanistan.[4] On October 7, the United States launched its first air strikes on Al Qaeda and the Taliban. The first two SOF teams flew into Afghanistan on October 19 and linked up with the Northern Alliance's military commander, General Fahim.[5] Two days later, they called in the first air strikes in support of the Northern Alliance's advance on the city of Mazar-e-Sharif.[6] The addition of SOF-directed precision air power greatly magnified the effectiveness of local forces. The result was a string of successes: the Northern Alliance took Mazar-e-Sharif on November 10, Kabul on November 13, and Kunduz on November 26. On December 6, two months after the start of the war, the Taliban leadership evacuated Kandahar, conceding defeat.

Like conflicts over the previous decade, the war in Afghanistan witnessed expert opinion that was more often wrong than right. Less than two weeks before the fall of Mazar-e-Sharif, for example, Harvard's Stephen M. Walt opined, "There's no evidence thus far that Taliban is unraveling as an administrative structure, and no sign as of right now that the Northern Alliance or anybody else is going to be able to defeat them militarily in the short term."[7]

The experts were particularly dubious that the combination of light ground forces and air power could be effective against the Taliban. The University of Chicago's John Mearsheimer argued, "American airpower is of limited use because there are few valuable targets to strike in an impoverished country like Afghanistan. Taliban ground forces are hard to locate and destroy from the air because, in the absence of a formidable ground opponent, they can easily disperse. Furthermore, the inevitable civilian casualties caused by the air assault are solidifying Taliban support within Afghanistan and eroding support elsewhere for the American cause."[8]

Mearsheimer instead argued that the United States needed to deploy a force of at least 500,000 troops to defeat the Taliban and crush Al Qaeda.[9] Mackubin Thomas Owens, for his part, recommended the deployment of two light infantry divisions and a corps headquarters but predicted that overthrowing the Taliban with even such a force would be daunting and would require either undertaking operations in the dead of winter or postponing an offensive until the spring of 2002.[10] Lawrence Kaplan likened the war in Afghanistan to "Clinton-era bombing campaigns in Bosnia and Iraq, in which air power was employed in modest increments and with exquisite calibration—bombing campaigns that not a few members of the Bush national security team at the time mocked at the time as feeble pinpricks."[11]

In the event, such prognostications proved wildly off target. In the course of two months, 316 SOF and 110 CIA paramilitary officers, working with native insurgents and backed by massive amounts of air power, brought the Taliban regime down and denied Al Qaeda the use of the country as a safe haven.[12] In fact, the small U.S. footprint was a key element of the strategy. Policymakers sought to avoid the sense of foreign occupation that had catalyzed Afghan resistance to the Soviet invasion in the 1980s.[13]

It is easy—too easy—to ridicule the pundits' predictions. Mearsheimer probably would have been correct in past wars, when precision-guided munitions (PGMs) were scarce and expensive. He also might have been correct if the United States had lacked the ability, provided by the combination of special operations forces and the Northern Alliance, to force Al Qaeda and Taliban forces to concentrate.

In fact, such predictions are most useful in helping us understand how the character of warfare was changing. Writing toward the end of the campaign, John Keegan forthrightly admitted, "Warfare is undergoing some strange transformations. Outcomes are becoming increasingly difficult to predict." He noted, "In the last 20 years, I have been required professionally to comment upon, to analyze, and to predict outcomes in five wars: The Falklands, the Gulf, the civil war in the former Yugoslavia, Kosovo and now Afghanistan. The task has become progressively more difficult."[14]

The "Afghan Model" of Warfare

To many, Operation Enduring Freedom introduced a new model of warfare, one composed of the networking of small, dispersed groups of SOF, indigenous forces, and precision air power.[15] The combination of SOF and indigenous forces and precision air power put Taliban and Al Qaeda forces on the horns of a dilemma: if they concentrated to repulse a ground attack, they became vulnerable to air strikes, but if they dispersed to reduce their exposure to air strikes, they opened themselves up to ground attack.[16]

Special Operations Forces

Special operations forces constituted one of the central ingredients of the Afghan model. For the most part, SOF units acted in their traditional unconventional warfare role by training, equipping, and leading native forces, with the latter doing the bulk of the fighting. But they also played another, less traditional, role, acting as sentient sensors to identify targets and direct air strikes, greatly increasing the lethality of precision-guided munitions. Nearly every Army Special Forces detachment or SEAL platoon included Air Force Special Operations Command combat controllers equipped with laser rangefinders, laser target designators, laptop computers, and ultra-high frequency (UHF) radios.[17] Air Force combat controllers used the AN/PEQ-1 SOF Laser Acquisition Marker (SOFLAM) laser rangefinder and designator. The system, which looked like a giant pair of elongated binoculars mounted on a small tripod, shot out a laser beam that allowed soldiers to calculate the coordinates of Taliban or Al Qaeda positions and call in air strikes.[18] Some combat controllers worked twenty-five days straight, calling in between ten and thirty attacks per day.[19] The results were often dramatic. For example, three SOF teams directed air strikes on Taliban and Al Qaeda positions during the 10th Mountain Division's assault on the Al Qaeda bastion in the Shah-i-Kot Valley, dubbed Operation Anaconda. According to one estimate, the thirteen commandoes in these teams were responsible for killing more enemy fighters than the remaining two thousand U.S. soldiers in the Shah-i-Kot valley combined.[20] During the Battle of Tora Bora, a team of three SOF and two CIA operatives infiltrated into the mountains housing Al Qaeda fighters and used laser designators to mark targets and call in air strikes for four days.[21]

Parts of the Defense Department had examined the use of small, dispersed ground forces to call in precision weapons for years. In 1996 the U.S. Defense Science Board sponsored a study of ways to increase the effectiveness of rapidly deployable units. The study's final report prefigured the problems confronting the United States in Afghanistan, as well as the constituent elements of the answer. Arguing that in many contingencies current U.S. forces would be too slow to arrive and vulnerable during deployment, the group called for a radical restructuring of a portion of the U.S. military, including the fielding of light, agile ground forces connected by a robust information grid and reliant upon remote sensors and weapons.[22]

The U.S. Marine Corps' 1996 Hunter Warrior experiment explored a similar concept. The exercise was designed to test the hypothesis that forces that move or mass on the battlefield could be targeted and destroyed by precise long-range fires. During the exercise, the Marines seeded the battlefield with dispersed rifle squads whose job it was to spot enemy forces and direct fires against them.[23] The squads, like the SOF units in Afghanistan half a decade later, were not combat formations per se, but in effect dispersed, sentient sensors.

The experiment received a harsh critique within the Marine Corps for being in opposition to the service's tradition of close combat and maneuver warfare. As one officer put it, "The Hunter Warrior concept is basically a technical concept for the efficient processing and coordination of fire support. It is essentially a procedure. Treated as an operating concept, it reduces practically the full art and science of war to the processing of targets.... The technology may be cutting-edge, but the operating concept is a direct descendant of the failed World War I French doctrine often referred to as 'methodical battle.'"[24]

Some critics argued that the concept of dispersed squads calling in remote fire reduced Marines to little more than human sensors serving high-technology machines. Others argued that it was at odds with the Marine Corps' tradition of amphibious warfare.[25]

The Marine Corps failed to follow up on the concepts explored in Hunter Warrior, turning attention instead to urban warfare. During Operation Enduring Freedom, Marine forces found themselves relegated to a supporting role. In the wake of the success of SOF in Afghanistan, some criticized the decision. Owen West, for example, argued that by abandoning the concepts examined in Hunter Warrior, the Marine Corps had ceded the initiative in developing innovative tactics to others.[26]

Human beings, not pieces of technology, stand at the heart of the culture of special operations. Special operations missions are ultimately about working with and motivating local forces. That having been said, a number of technologies allowed small SOF teams to be so effective. Night vision goggles, or NVGs, gave U.S. forces an advantage over their adversaries. The goggles were the descendants of the Vietnam-era Starlight scope, which amplified existing light to allow marksmen to identify targets at night.[27] Conventional forces in Afghanistan used the binocular AN/PVS-7 NVG, and SOF used the lighter and more advanced monocular AN/PVS-14.[28] NVGs allowed the U.S. armed forces to dominate adversaries who, lacking the ability to see in the dark, could not mount offensives at night. Indeed, the U.S. armed forces' widespread use of NVGs in recent wars has reversed a historical pattern: traditionally, the weaker side preferred to fight at night in order to nullify the superiority of the stronger side. Now the technologically superior side owns the night.

Precision-Guided Munitions

Another element of the "Afghan model" was the widespread, even routine, use of PGMs, which multiplied the firepower available to U.S. and Afghan forces. American soldiers were able to call down precision weapons not just on Taliban tanks and armored vehicles, but also fortifications and even troop concentrations. In other words, PGMs began to substitute not only for unguided bombs, but also for artillery and other forms of fire support. U.S. air power also raised the morale and stiffened the resolve of Afghan guerrillas. Conversely, when air strikes failed to materialize, such as at the beginning of Operation Anaconda, Afghan fighters became demoralized.[29]

The war continued the trend toward the increasing use of PGMs. Whereas only 9 percent of the munitions expended during the 1991 Gulf War featured precision guidance and 29 percent of the munitions employed during the 1999 air war over Kosovo were guided, during Operation Enduring Freedom two years later nearly 60 percent were guided.[30]

Operation Enduring Freedom saw the widespread use of the global positioning system (GPS) guided Joint Direct Attack Munition, or JDAM. First employed in Kosovo, JDAM became the weapon of choice of U.S. forces in Afghanistan. Between October 2001 and February 2002, U.S. forces dropped 6,600 of the munitions, and during a ten-minute period on October 18, 2001, U.S. air forces dropped a hundred of the bombs.[31]

Given Afghanistan's remoteness, bombers played a large role in the conflict. At the beginning of the campaign, they were used in the conventional roles of destroying the small Taliban air force and air defenses and disrupting command and control systems. Beginning October 21, however, they began to fly missions in support of the Northern Alliance. Loitering high over the battlefield for hours, bombers were guided by ground-terminal attack controllers as they dropped PGMs against battlefield targets. Often, aircraft delivered their ordnance within minutes of a request. Between October 2001 and March 2002, ten B-52s and eight B-1s out of Diego Garcia, an island in the Indian Ocean, and B-2s out of Whiteman AFB, Missouri, flew more than 48 percent of combat missions in Afghanistan and dropped nearly seven thousand tons of munitions, approximately 75 percent of the campaign's total.[32]

The widespread employment of GPS-guided munitions like JDAM changed the role of the aircrew in combat. Aircrews dropping laser-guided bombs such as those used in the 1991 Gulf War had to identify their target, designate it with a laser, and keep the designator on the target throughout the bomb's flight. Those dropping JDAMs, by contrast, merely had to approach to within the weapon's effective envelope and drop the weapon. Although pilots still needed skill to avoid defenses on the way to the target, they took much less of an active role in bombing itself.

Although PGMs proved to be discriminate, they were not foolproof. On December 5, U.S. forces suffered their worst friendly-fire incident of the war when an air controller confused the GPS coordinates of his team

for those of the target and called a 2,000-pound bomb onto his position. Three U.S. soldiers and five Afghan militiamen were killed. Doz-
ens more, including the future president of Afghanistan, Hamid Karzai,
were wounded.[33]

Not all the aircraft providing support were so modern. One of the fa-
vorite aircraft of SOF commanders was the venerable AC-130U Spectre,
with its optical and infrared sensors, as well as its 25mm gun, 40mm
cannon, and 105mm cannon. Although originally used in Vietnam, the
Spectre had been modernized repeatedly over the years. It put a combination of advanced sensors and large volumes of precise firepower
at the disposal of ground forces. The aircraft could orbit overhead for
hours, identifying and destroying enemy formations. The Air Force also
established real-time links between Predator unmanned aerial vehicles
(UAVs) and AC-130 Spectre gunships. As a result, when UAVs were able
to spot concentrations of enemy fighters, an AC-130 would often arrive
within minutes.

The AC-130 put awesome firepower in the hands of its operators. As
with all weapons, however, it was subject to human error, as when an
AC-130 mistook a group of U.S.-led Afghan fighters for the enemy during
the opening phases of Operation Anaconda in the Shah-i-Kot valley.[34] In
July 2002, an AC-130 mistook fire from a wedding party for an attack and
opened fire, reportedly killing 48 and wounding 117.[35]

Unmanned Aerial Vehicles

The war in Afghanistan also saw the widespread use of UAVs and the combat debut of unmanned combat air vehicles, or UCAVs. UAVs such as the
RQ-1 Predator and the RQ-4 Global Hawk gave coalition forces the ability
to "stare" at the battlefield for hours at a time, something that neither satellites nor high-flying, fast-moving reconnaissance aircraft could do. The
high-altitude, long-endurance Global Hawk, which had its combat debut
in the skies over Afghanistan, could loiter at sixty thousand feet—above
the altitude of most air defenses—for eighteen hours.

The Predator first flew over Afghanistan more than a year before Operation Enduring Freedom. The CIA flew Predators over Afghanistan ten
times beginning in 2000 in an effort to identify the whereabouts of Usama
bin Laden. On two occasions, analysts believed that they spotted bin Lad-
en. However, they lacked the ability to strike him.[36]

Fortunately, the Defense Department already had an effort underway to
arm the Predator. The Air Force launched a program to field a version of
the UAV armed with two laser-guided AGM-114C Hellfire air-to-surface

missiles in 2000, and the program proceeded at an extraordinary pace. Before September 11, however, its employment remained controversial. The Air Force and CIA argued over who would bear the cost of the program, who would control the aircraft, and who would have the authority to fire its weapons.[37]

If before September 11 nobody wanted control of (and responsibility for) the armed Predator, after the attacks everyone wanted it. During Operation Enduring Freedom, Predators fired 115 Hellfire missiles and used their lasers to designate 525 targets for attack by manned aircraft.[38] In some cases, they were used to deliver ordnance in extremely difficult circumstances, as when a Predator delivered a Hellfire missile against a target in support of a Special Forces team that was under fire during Operation Anaconda.[39] The notion of an unmanned vehicle controlled by an operator located hundreds or thousands of miles away delivering bombs in support of troops in close combat is something that would have previously been inconceivable. Indeed, the pitfalls of such an approach were apparent in another incident in which a Predator nearly bombed a CIA team that its operators mistook for the Taliban.[40]

The availability of communications bandwidth proved to be a major constraint on the use of UAVs. During 2001 and 2002, for example, the Air Force was able to keep only two Predators and one Global Hawk operational over Afghanistan simultaneously. A single Global Hawk consumed about five times the total bandwidth consumed by the entire U.S. military during the 1991 Gulf War.[41]

The armed Predator continues to be used to hunt down and kill terrorists. In November 2002, a Hellfire launched by a CIA-controlled Predator destroyed a car carrying six terrorists, including Salim Sinan al-Harethi, Al Qaeda's chief operative in Yemen and a suspect in the October 2000 bombing of the destroyer USS *Cole*. In May 2005, a Predator-launched missile killed another Al Qaeda leader, Haitham al-Yemeni, in Pakistan.[42] And in January 2006, a Predator killed a number of Al Qaeda operatives in an attempt to kill the network's second-in-command, Ayman al Zawahiri.[43]

The widespread ability to collect and distribute surveillance information quickly had a number of unintended consequences. First, it created a dependency on—some might say an addiction to—reconnaissance and surveillance data. As one operator put it, "The special ops community has gotten so that we can't go in now unless a UAV is looking at it or an AC-130 is looking at it."[44]

Second, at the highest levels, the combination of networking and real-time information sources fostered an odd mixture of disengagement and micromanagement by the CENTCOM leadership. On the one hand, General Tommy Franks ran the war from his headquarters in Tampa, Florida, eight time zones away from Afghanistan. On the other hand, many officers decried what they saw as the command's "command-by-video teleconference" mindset.[45]

The accessibility of global communications exacerbated the trend toward centralized military operations. It produced a shared picture of the battlefield at all levels and permitted "command at a distance." Not only did it enable more timely operations, but it also allowed senior leaders to become enmeshed in the minutiae of operational planning and execution.

The availability of Predator video in headquarters at times proved to be a distraction or even an irresistible temptation to micromanagement.[46] In his autobiography, General Tommy Franks, the commander of U.S. Central Command during Operation Enduring Freedom, recounts—entirely without irony—an incident in which he, the commander of coalition forces in Afghanistan, sat for hours watching Predator video of a convoy believed to be carrying Mullah Omar, the Taliban leader.[47] Franks, the senior commander, sat—his trusty lawyer at his elbow—and barked orders usually given by a junior officer. As he recounted:

> "Give me an exact fuel status on the UAV," I ordered the operator.
> "Sir, we're good for two hours, ten minutes before Bingo fuel and [Return to Base]...."
> "Hold station," I ordered.
> Several people from the convoy ran into a compound on the right side of a street. "How long to orient the Predator and take a Hellfire shot?" I asked.
> "Valid target for Hellfire," [Franks's lawyer] added as we waited for the answer.
> "Lining up for a shot," the operator at Langley said. "About five minutes until launch."

Franks eventually authorized a strike on a house that he believed contained enemy leaders, but they were not inside.

To compound the irony of the incident, Franks recounts receiving a phone call from the chairman of the Joint Chiefs of Staff, Air Force General Richard Myers, informing him that Air Force Chief of Staff General John Jumper had been watching the same Predator video and had noticed that the Taliban leaders had left the house before the Hellfire had struck it.[48] In other words, the video broadcast has transfixed not one but three four-star generals.

The thought that a general with thirty-six years of military experience would himself command an aircraft to drop a bomb on a house would previously have been inconceivable. With a manned aircraft, a junior officer would have performed the same function. The availability of real-time imagery, however, encouraged interference.

The Legacy of Afghanistan

The defeat of the Taliban three months after the September 11, 2001, terrorist attacks was a visible setback for Al Qaeda. Moreover, it defied prewar predictions that the campaign would turn into a bloody quagmire. In the event, the United States and its allies overthrew the Taliban and denied Al Qaeda a sanctuary at the cost of only thirty-nine dead, just sixteen of whom died in combat.[49]

The Afghan model became quite influential. Some, such as Jeffrey Record, argued that it resonated with American strategic culture. As he wrote, "An airpower-dominant way of war in which U.S. ground forces ... are ancillary, functioning mainly as target spotters and liaisons to indigenous proxies, is an inherently attractive way of war, especially for a society that values the individual as highly as America's does. It also permits a casualty-phobic political and military leadership to wage war effectively—i.e., to achieve decisive strategic effects without paying the blood price traditionally associated with attainment of those effects."[50]

Others, such as Vice Admiral Arthur K. Cebrowski, the Pentagon's Director of the Office of Force Transformation, argued that it amounted to an "emerging American way of war."[51] As he put it in October 2002:

> We are entering a new era of military operations and capabilities. The very character of warfare is changing to account for the massive implications of the information age. It embodies the new decision logic with attributes we will become increasingly familiar with and comfortable. We can already see its effects in current operations. The last time we witnessed change of this magnitude was with the advent of the industrial age and the levée en masse ... Both of these events are rapidly receding into the past. A new American way of war has emerged—network-centric operations.[52]

President Bush used the success of U.S. forces in Afghanistan to give defense transformation a new boost. As he told the corps of cadets at The Citadel military college on December 11, 2001, "The conflict in Afghanistan has taught us more about the future of our military than a decade of blue ribbon panels and think-tank symposiums." He saw the experience of Afghanistan as directly relevant to the future. As he put it, "We're entering an era in which unmanned vehicles of all kinds will take on greater importance—in space, on land, in the air, and at sea. Precision-guided munitions also offer great promise."[53]

The president argued that the need to defeat Islamic terrorist networks would spur innovation in the U.S. armed forces. In his words, "Our military culture must reward new thinking, innovation, and experimentation.... And every service and every constituency of our military must be willing to sacrifice some of their own pet projects. Our war on terror cannot be used to justify obsolete bases, obsolete programs, or obsolete weapon systems. Every dollar of defense spending must meet a single test: It must help us build the decisive power we will need to win the wars of the future."[54]

Not all observers shared the president's view of the war. Critics argued that the outcome of the war was idiosyncratic. They contended that it would be foolish to rely on light ground forces backed by air power in future conflicts. For example, as the U.S. Army War College's Stephen Biddle wrote:

> The actual fighting in Afghanistan involved substantial close combat.... In Afghanistan, U.S. proxies with American support brushed aside unskilled, ill-motivated Afghan Taliban, but against hard-core al Qaeda opposition, outcomes were often in doubt even with the benefit of 21st century U.S. air power and American commanders to direct it.... We should be wary of suggestions that precision weapons, with or without special operations forces to direct them, have so revolutionized warfare that traditional ground forces are now superseded.[55]

Biddle argued, for example, that the failure to commit properly trained and motivated ground troops at Tora Bora allowed Al Qaeda's leadership to escape.

In the end, the Afghan model was to prove less influential within the U.S. armed forces than its backers hoped. The U.S. war in Iraq gave general-purpose forces a more central role than the campaign in Afghanistan had. In its own way, however, it demonstrated just how much the U.S. military had changed since 1991.

Iraq, Part One

Less than two years after the onset of Operation Enduring Freedom, the United States once again went to war, this time to overthrow Saddam Hussein's Ba'athist regime in Iraq. The reasons for the war were many, including Saddam Hussein's steadfast refusal to comply with a host of United

Nations resolutions, the continued threat he posed to his neighbors, and his systematic efforts to terrorize his own population. Foremost in many people's minds was concern that Iraq possessed chemical and possibly biological weapons and had reconstituted its nuclear weapons program. The alternative to overthrowing Saddam Hussein's regime, containment, was losing its viability as countries such as Russia and France sought to loosen or lift sanctions on Iraq. Less discussed, but perhaps more important, was the fact that sanctions and the presence of large numbers of U.S. forces in Saudi Arabia, needed to enforce the no-fly-zone over southern Iraq, had become a rallying cry for jihadists such as Usama bin Laden.[56]

Operation Iraqi Freedom offered a contrast with Operation Desert Storm, its predecessor twelve years earlier. The United States waged the 1991 Gulf War under a doctrine of overwhelming force that placed a premium on holding risks to an absolute minimum. It began after the methodical buildup of combat power in the theater. The war itself started with a thirty-nine-day air campaign; a ground force of over 500,000 delivered the coup de grace in a campaign lasting one hundred hours. In contrast to its predecessor, Operation Iraqi Freedom emphasized surprise in an effort to prevent Iraq from sabotaging its oil infrastructure and using its presumed chemical and biological weapons against coalition forces. At 180,000, the ground component was less than two-fifths the size of coalition forces in 1991.

The enemy the United States faced in 2003 also offered a contrast with that of 1991. In 1991, Iraq had had 950,000 troops organized into sixty divisions; the Republican Guards alone had numbered 150,000. Moreover, Baghdad possessed more than 5,000 tanks, 5,000 armored personnel carriers (APCs), and 3,000 artillery pieces. In 2003, by contrast, the Iraqi army totaled between 280,000 and 350,000 troops in seventeen divisions, and the Republican Guard had shrunk to 80,000. Baghdad's arsenal had diminished to some 2,200 tanks, 2,400 APCs, and 4,000 artillery pieces.[57]

As had occurred prior to and during the campaign in Afghanistan, a host of analysts offered their forecasts of the course of the war in Iraq. Among the most pessimistic were some of the leading commanders of the last war with Iraq. Given their military expertise, their views deserved to be taken seriously. The fact that their predictions proved so wide of the mark provides further evidence of the changing character of war.

Several prominent retired officers argued at the outset of the war that U.S. forces were too weak to defeat the Iraqi army. Retired General Barry McCaffrey portrayed the decisive match up of the war as that between five Iraqi armored divisions, on the one hand, and one U.S. armored division supported by the "modest" armor of the 1st Marine Division and the 101st

Airborne Division's AH-64 Apache attack helicopters on the other. In his view, the United States was in a weak position: "We should be fighting this battle with three U.S. armored divisions and an armored cavalry regiment to provide rear area security. We also have inadequate tube and rocket artillery to provide needed suppressive fires for the joint team." As he saw it, the United States was "overextended and at risk." To enjoy success, McCaffrey recommended that the president call up at least three Army National Guard divisions for thirty-six months and significant numbers of Marine Corps, Navy, Coast Guard, and Air Force reserve units.[58]

McCaffrey's argument typifies a view of warfare rooted in World War II, developed in the Cold War, and expressed in the first Gulf War. In that view, heavy armored and mechanized forces were the sine qua non of land warfare; light forces were subordinate. Similarly, air power could not substitute for artillery, let alone heavy forces.

When Secretary of Defense Rumsfeld questioned McCaffrey's judgment, McCaffrey responded: "I'm a professor of national security studies, and I know a lot more about fighting than [the Secretary of Defense] does.... The problem isn't that the V Corps serving officers are commenting or that retired senior officers are commenting on television. The problem is that they chose to attack 250 miles into Iraq with one armored division and no rear-area security and no second front."[59]

The record of the pundits such as McCaffrey shows that personal bravery and military expertise do not equate to strategic insight, and that the lessons of past wars can mislead as well as inform. If the Gulf War veterans had gotten their way—more heavy divisions, necessitating a long buildup—the result likely would have been more time for the Iraqi government to prepare for attack, greater destruction of Iraq's infrastructure, and more civilian casualties. Nor would such formations have been equipped or trained to deal with the insurgency that followed.

The war began a day and a half earlier than planned after the United States received intelligence indicating that Saddam Hussein, his sons, and Ba'athist leaders were meeting at a compound in the Dora Farms neighborhood of Baghdad. As a result, President Bush authorized a strike by Tomahawk cruise missiles and a pair of F-117 aircraft equipped with EGBU-27 laser-guided bombs.[60] The weapons were precise; the intelligence that produced their targets was not. In fact, Saddam and his sons were nowhere near the site. A report later in the war that Saddam and his sons were meeting at a house in the Al-Mansur district of Baghdad that triggered another air strike was similarly wrong. In fact, not one of the top two hundred Iraqi leaders was killed by an air strike.[61] The war served as a vivid demonstration of the continuing relevance of what Clausewitz termed "friction."

The ground attack was carried out by four bulked-up divisions—the U.S. Army's 3rd Infantry Division and 101st Airborne Division, the U.S. Marine Corps' I Marine Expeditionary Force, and the 1st United Kingdom Division. The Army forces formed the main effort, dashing west of the Euphrates River to seize Baghdad. The Marines advanced east of the Euphrates, pinning down the Iraqi army and drawing fire away from the main effort. The British, for their part, invested and seized the port city of Basra, Iraq's second largest city. U.S. troops reached Baghdad in sixteen days and occupied Saddam Hussein's hometown of Tikrit in twenty-seven.

SOF played an important role in the Iraq War. Although the Pentagon's leadership rejected efforts to apply the "Afghan model" to Iraq and rely on SOF, air power and indigenous forces to do the bulk of the fighting, SOF played a more prominent role in Iraq in 2003 than they had in 1991. In the south, Navy SEALs seized Iraq's oil export infrastructure, preventing Saddam Hussein's regime from destroying it. In the north, in a limited application of the "Afghan model," SF and Kurdish *pesh merga* militia pinned down 40 percent of Iraqi divisions.[62] In the west, SOF seized Iraqi military facilities in an attempt to deny Baghdad the ability to launch missiles against Israel and the coalition.[63]

Coalition forces faced some difficult days, but none worse than March 23. On that day, the Marine Corps' Task Force Tarawa fought a sharp engagement in An Nasiriyah, losing eighteen. That same day, the Army's 507th Maintenance Company drove into an ambush there, resulting in eleven killed, nine wounded, and seven captured. Farther north, the Army's 11th Attack Helicopter Regiment launched an unsuccessful deep attack against the Iraqi Republican Guards' Medina Division near Karbala. The operation, which caused minimum damage to the Iraqis, damaged literally every one of the attacking AH-64 Apache helicopters.[64]

Despite such setbacks, the major combat phase of the war proved to be nearly bloodless for the United States and its allies. By May 1, U.S. and British forces had lost a total of 169 killed, the lowest daily casualty rate of any conflict since the Revolutionary War.[65] In John Keegan's judgment, it was "a collapse, not a war." The seeming ease of the victory spawned the strange phenomenon of a western commentator helpfully explaining to their audiences how the Iraqis *should* have waged the war, as if a war was not worth winning unless preceded by blood and suffering.[66]

Precision-Guided Munitions

The Iraq War demonstrated the continuing U.S. dominance of the air. Coalition aircraft flew just over 41,400 sorties during the period of major combat operations from March 19 to April 30. Only one fixed-wing aircraft—an A-10A—and six helicopters were lost to enemy fire.[67]

The war also continued the trend toward the use of PGMs. Whereas 8 percent of the munitions employed in the 1991 Gulf War were guided, 68 percent of those employed in the 2003 Iraq War were guided. These included 802 BGM-109 TLAM and 153 AGM-86C/D CALCM. By far the most common PGM, however, was the JDAM; coalition air forces dropped than 6,500 of the weapons in several varieties.[68]

Coalition precision air attacks broke the will of the Iraqi defenders. As Lieutenant General Majid Husayn Ali Ibrahim Al-Dulaymi, commander of the Iraqi Republican Guard I Corps, said after the war, "Our units were unable to execute anything due to worries induced by psychological warfare. They were fearful of modern war, pin-point war in all climates and in all weather." He recalled visiting the Adnan Republican Guard Division shortly after precision air attacks had destroyed one of its battalions. As he put it, "The level of precision of those attacks put real fear into the soldiers of the rest of the division. The Americans were able to induce fear through the army by using precision air power."[69]

Whereas clouds and dust interfered with laser-guided bombs in the 1991 Gulf War, the GPS-guided weapons suffered no such limits. Between March 25 and 28, Kuwait and southern Iraq were blanketed in a blinding sandstorm. Iraqi forces hunkered down, unable to maneuver, safe in the belief that they could not be seen. U.S. aircraft, however, used GPS-guided weapons to decimate Iraqi units.[70]

The Iraqis developed countermeasures to U.S. air power. For example, they stationed inoperable equipment where it could be seen in order to deceive U.S. aircrews. They hid units in palm groves or near no-strike zones like hospitals and schools. They used cell phones, low-power radios, and couriers to communicate.[71] They also attempted to counter U.S. PGMs by deploying devices to jam their GPS signals. Such countermeasures proved ineffective, however. According to U.S. officers, U.S. forces were able to identify the jammers and destroy them with EGBU-27 PGMs.[72]

GPS data not only helped guide U.S. bombs, but it also allowed commanders to locate their forces on the battlefield. The Army's system, dubbed Force XXI Battle Command Brigade and Below (FBCB2) but known more widely as Blue Force Tracking (BFT), used GPS transponders to transmit a coded signal containing its coordinates, direction, and speed via satellite to each headquarters, to CENTCOM, and to the Pentagon. Commanders at these locations were able to track friendly forces down to the level of a tank company or individual aircraft.[73]

The first units to experiment with the system in the mid-1990s found that it improved positional reporting and command and control. As a result, the Army designated the 4th Infantry Division as the first division to be equipped with the system. Its soldiers learned that it allowed them to operate more quickly and accurately with improved situational awareness.[74]

Deployment of the equipment to other land forces began in late 2002 and continued up to the start of the war. Over a three-month period before the war, the Army and its contractor installed available equipment on vehicles and helicopters and trained commanders and soldiers in its use. By the time war came, more than 1,200 systems were installed with Army, Marine, and British forces.[75]

The ability to track friendly forces improved command and control of coalition forces. According to Major General Buford Blount, the commander of the 3rd Infantry Division, BFT gave him the ability to control his division over a front of between two hundred and three hundred kilometers, ten times the frontage the same division had occupied in the 1991 Gulf War. The technology also accelerated the pace of operations and reduced drastically fratricide. During the major combat phase of Operation Iraqi Freedom, only one soldier was killed by friendly direct ground fire. This is a stark contrast to Operation Desert Storm, during which thirty-five were killed and seventy-two were wounded by friendly fire.[76]

Soldiers also used the digital systems in innovative ways. Many commanders relied heavily on the FBCB2 system's email and chat-room capability, even though they were designed for administrative use in garrison rather than as a means of battlefield communication. Soldiers also found that they could use the system to navigate in zero-visibility conditions, a function never foreseen by its developers.[77]

Another notable contrast was the effectiveness of theater missile defense. Whereas U.S. theater ballistic missile defenses during the 1991 Gulf War were marginally effective, in 2003 U.S. and Kuwaiti Patriot theater ballistic missile defense (TBMD) batteries intercepted and destroyed all nine Iraqi ballistic missiles launched at military targets. This effectiveness was due to the deployment of more effective interceptors, particularly the Patriot Advanced Capability (PAC)-2 Guidance Enhanced Missile and PAC-3 interceptors. When the war started, there were twenty-seven U.S. and five coalition Patriot batteries in Kuwait, Jordan, Qatar, Bahrain, and Saudi Arabia, with additional units in Israel and Turkey. Also important was the establishment of networks that linked Patriot batteries with the AEGIS destroyer USS *Higgins*, which provided early warning of ballistic missile attack.[78]

U.S. missile defenses proved far less effective against cruise missiles. Indeed, they failed to detect or intercept any of the five HY-2/CSSC-3 Seersucker cruise missiles launched against Kuwait. One came close to hitting Camp Commando, the U.S. Marine Corps headquarters in Kuwait, on the first day of the war. Another landed just outside a shopping mall in Kuwait City. The missiles also contributed to fratricide that caused the loss of two coalition aircraft and the death of three crewmembers.[79]

Networked Operations

The war in Iraq demonstrated the expanding use of information networks by U.S. forces. CENTCOM fought the war from command posts in four countries: Franks from Al Udeid airbase in Qatar; his air commander, Lieutenant General Michael Mosely, from Prince Sultan Air Base in Saudi Arabia; his naval commander, Vice Admiral Timothy Keating, from Bahrain; and his ground commander, Lieutenant General Michael McKiernan, in Kuwait. U.S. commanders enjoyed forty-two times as much bandwidth as their counterparts in the 1991 Gulf War.[80] They also benefited from the Secret Internet Protocol Router Network, or SIPRNET, a classified version of the Internet that allowed planners and leaders to collaborate and gave them desktop access to the latest plans and intelligence, secure email, and chat rooms.[81]

The existence of such information systems did not, however, lift the fog of war from the battlefield, as some defense analysts had predicted in the 1990s. Although commanders at the division level and above often had an excellent view of the battlefield, troops at the brigade and battalion received much less information. Moreover, communication systems proved inadequate, due to the speed of advance of U.S. forces over long distances. Fast-moving forces tended to outrun their communication links to higher headquarters, and signal units were unable to provide support to the U.S. Army at the lowest levels.[82]

The largest Iraqi counterattack of the war, which occurred early on April 3 near a key Euphrates River bridge about thirty kilometers southwest of Baghdad, surprised U.S. forces. U.S. sensors failed to detect the approach of three Iraqi brigades. The ensuing battle pitted the 3rd Infantry Division's Task Force 3–69 against a total of eight thousand soldiers backed by seventy tanks and armored personnel carriers.[83] Even though it was outnumbered and lacked intelligence on the enemy it faced, the U.S. force prevailed after a difficult battle.

One of the best-known episodes of the war was the 2nd Brigade, 3rd Infantry Division's "Thunder Run" up Highway 8 and then into Baghdad in the first week of April 2003. While the maneuver was a tactical and operational success, it took place in an environment of great uncertainty over the location and intentions of Iraqi forces. U.S. commanders were unable to get accurate, detailed information about Iraqi strength and dispositions or even to find out whether SOF had scouted the area.[84] In two instances, U.S. forces took wrong turns on an interstate highway.

Fortunately for the United States, the Iraqi situation was worse. As David Zucchino wrote, "Senior Iraqi officers in the capital seemed content to believe their own lies, that the war was going well and the Americans were bogged down south of the city. Even many ordinary civilians seemed unaware that there was a war going on. Despite the columns of black smoke from burning vehicles and the thunderous pounding of U.S. tanks and the Bradleys, civilians in family sedans were [driving] ... like it was just another Saturday morning in the suburbs."[85]

In one particularly telling incident, Republican Guard General Sufian Tikriti was killed when he drove into a Marine Corps blockade.[86] Apparently he and his guards did not know that the Marines were there.[87]

The swift and lopsided outcome of the war in Iraq provided additional evidence of a change in the character of war. As President Bush declared in a speech soon after the fall of Baghdad, "By a combination of creative strategies and advanced technology, we are redefining war on our terms ... more than ever before, the precision of our technology is protecting the lives of our soldiers, and the lives of innocent civilians.... In this new era of warfare, we can target a regime, not a nation."[88] Max Boot was even more enthusiastic, labeled the war "one of the signal achievements in military history." As he saw it, "Spurred by dramatic advances in information technology, the U.S. military has adopted a new style of warfare that eschews the bloody slogging matches of old. It seeks a quick victory with minimal casualties on both sides. Its hallmarks are speed, maneuver, flexibility, and surprise. It is heavily reliant upon precision firepower, Special Forces, and psychological operations. And it strives to integrate naval, air, and land power into a seamless whole."[89]

Iraq, Part Two

Such exhilaration proved to be short-lived, as triumph in conventional operations yielded to the indecisiveness of occupation and counterinsurgency. The period following "major combat operations" proved to be far deadlier than that which preceded it. From March 19, 2003, to April 16, 2006, the United States suffered 3,773 fatalities, including 3,086 to hostile action.

Thirty-nine percent of fatalities were from improvised explosive devices (IEDs), 31 percent from other hostile fire, 4 percent from car bombs, and 3 percent from mortars and rocket-propelled grenades (RPGs).[90]

One prominent British critic of the U.S. conduct of counterinsurgency in Iraq has singled out the American military's reliance on advanced technology as a cause of its difficulty in waging irregular warfare. As he has written, "The lure of technology can be misleading. In an environment where, above all else, it is imperative that the occupying force be seen as a force for the good, it is counter-productive when technological solutions are employed that promote separation from the population. Furthermore, a predilection with technology arguably encourages the search for the quick, conventional solution, often at the expense of the less obvious, but ultimately more enduring one."[91]

Technology clearly offers no silver bullets for countering insurgents. That is not to say, however, that technology is irrelevant. It has, for example, provided the United States the ability to protect soldiers against the weapons of the insurgents, including body armor, armored vehicles, and technologies to counter IEDs.

The spectacle of U.S. forces suffering casualties while riding in unarmored vehicles led to efforts to improve the protection afforded to U.S. vehicles. Of primary concern was the High Mobility Multipurpose Wheeled Vehicle (HMMWV, or "Humvee"), a utility vehicle designed for service behind the front line. In an insurgency, however, there are not front lines. Soldiers traveling in vehicles never meant for the front lines now faced suicide bombers, RPGs, and IEDs. As a result, the Defense Department launched a series of programs to provide supplemental armor for vehicles in the field. Individual units also improvised armor, including scrap metal and sandbags.

As of July 2005, two-thirds of the Humvees in Iraq had been factory armored with an additional two thousand pounds of protective plating and equipped with bulletproof windows, a configuration dubbed the M-1114. Such modifications provide protection against AK-47 assault rifle fire and antitank mine blasts. U.S. forces also began deploying the M-1117 Armored Security Vehicle. Although the vehicle provides greater protection than the M-1114, it costs more than five times as much.[92]

Insurgents responded to the up-armoring of U.S. vehicles with mass, fielding bigger and more sophisticated bombs. The balance between protection and raw explosive power greatly favored the latter. Indeed, some of the insurgents' explosive devices were large enough to penetrate virtually any thickness of armor. On July 23, 2005, for example, a 500-pound bomb detonated underneath a Humvee near Baghdad, killing all four passengers and leaving a crater six feet deep and seventeen feet wide. On August 3, an IED flipped a twenty-five-ton amphibious assault vehicle, killing all fourteen Marines inside. The insurgents also fielded more effective IED designs, including shaped charges, with assistance from Iran.[93]

Nothing illustrates the lethality of IEDs better than the damage the most heavily armored of U.S. vehicles, the M-1 Abrams tank, sustained. During the 1991 Gulf War, eighteen M-1s were knocked out of action but no soldiers were killed. Between March 2003 and March 2005, eighty of the tanks were badly damaged and 70 percent were struck by enemy fire.[94]

The widespread use of IEDs has led to a broad-based effort to counter them. U.S. forces are, for example, reportedly employing the Warlock radio frequency (RF) jammer to intercept the signal sent from a remote location to the IED instructing it to detonate. Hundreds of the devices have been sent to Iraq.[95] U.S. forces are also using UAVs with electro-optical sensors to spot teams emplacing IEDs and radars to spot changes in the landscape indicating the presence of IEDs.[96] The Army is also operating C-12 Horned Owl aircraft equipped with radar and infrared sensors. The aircraft, which fly at night, monitor activity on the ground. Its sensors have the ability to detect objects buried in the ground as well as detect changes in ground features. The Air Force has also reportedly been using EC-130 Compass Call aircraft to jam remote triggering devices.[97]

The war in Iraq also illustrated the effectiveness of modern body armor. Most soldiers in Iraq wear armor made of ceramic plates embedded in Kevlar. These vests are lighter, more flexible, and vastly more protective than previous models. Whereas previous "flak jackets" provided protection against shrapnel, ceramic body armor could protect against automatic rifle fire. During the battle of Tora Bora, for example, a Taliban fighter shot a Special Forces soldier with three AK-47 rounds to the chest at close range. The soldier dropped to the ground for a few moments before getting back up and shooting his attacker dead.[98] Current Kevlar helmets can stop a pistol round, whereas their predecessors only provided protection against shrapnel.

The combination of improved body armor, advances in medicine, and faster airlifting of the wounded has saved many lives in Iraq. A soldier who arrives alive at a field hospital has a 96 percent chance of survival. On the other hand, saving lives has also led to more amputations, blinding, and brain damage. In Iraq, there were ten soldiers wounded for each one killed, more than double the rate in Korea, Vietnam, and the Gulf War.[99] However, the widespread use of IEDs and rocket-propelled grenades by insurgents led to many more serious wounds. Although body armor shields the soldier's chest, it leaves his or her extremities vulnerable. Indeed, nearly half of all troops wounded in Iraq since the fall of Saddam have been hit in their lower extremities, and a quarter have been hit in the hand or arm. The number of soldiers who have undergone amputations is double that of past conflicts. Moreover, nearly a quarter of wounded troops have suffered from traumatic head injuries and brain damage.[100]

The Global War on Terrorism illustrates both the utility and the limitations of advanced technology. Advanced military technology helped the United States achieve quick decisive victories in Afghanistan and Iraq. It did not, however, offer a panacea for insurgency.

Advanced technology was far from useless in combating the Iraqi insurgency, just as it had been far from useless in Vietnam. To take but a single case, the fact that U.S. troops in Iraq possess body armor capable of protecting them from automatic rifle fire has saved numerous lives. Moreover, to the extent that combat deaths erode public support, body armor permits the United States to stay the course in a protracted counterinsurgency struggle.

Notes

1. George W. Bush, "Address to a Joint Session of Congress and the American People," September 20, 2001, at www.whitehouse.gov/news/releases/2001/09/print/20010920-8.html (accessed April 18, 2006).

2. Ibid.

3. On the discussion of options for Afghanistan, see Bob Woodward, *Bush at War* (New York: Simon & Schuster, 2002), chapter 6.

4. Gary Schroen, *First In* (New York: Ballantine Books, 2005), 87.

5. Linda Robinson, *Masters of Chaos* (New York: Public Affairs, 2004), 156.

6. David R. Brooks et al., *The First Year: U.S. Army Forces Central Command During Operation Enduring Freedom* (Carlisle, PA: U.S. Army War College, 2002), 30.

7. Patricia Cohen, "Getting It Right: Strategy Angst," *New York Times*, October 27, 2001.

8. John J. Mearsheimer, "Guns Won't Win the Afghan War," *New York Times*, November 4, 2001.

9. Ann Scott Tyson, "Talk Grows of a Major U.S. Troop Deployment," *Christian Science Monitor*, November 1, 2001.

10. Mackubin Thomas Owens, "How to Win: The Case for Ground Troops," *Wall Street Journal*, October 31, 2001.

11. Lawrence F. Kaplan, "Ours to Lose," *New Republic*, November 12, 2001.

12. Benjamin S. Lambeth, *Air Power Against Terror: America's Conduct of Operation Enduring Freedom* (Santa Monica, CA: Rand Corporation, 2005), 160–61.

13. Henry A. Crumpton, "Intelligence and War: Afghanistan, 2001–2002," in *Transforming U.S. Intelligence*, ed. Jennifer E. Sims and Burton Gerber (Washington, DC: Georgetown University Press, 2005), 162–79.

14. John Keegan, "The Changing Face of War," *Wall Street Journal Europe*, November 26, 2001.

15. "Counter-Terrorism and Military Transformation: The Impact of the Afghan Model," in International Institute for Strategic Studies, *Strategic Survey 2002/3* (Oxford: Oxford University Press, 2003).

16. Richard B, Andres, Craig Wills, and Thomas Griffith Jr., "Winning with Allies: The Strategic Value of the Afghan Model," *International Security* 30, no. 3 (Winter 2005/2006): 124–60. See also Stephen D. Biddle, "Allies, Airpower, and Modern Warfare: The Afghan Model in Afghanistan and Iraq," ibid., 161–76.

17. Lambeth, *Air Power Against Terror*, 259.

18. Vernon Loeb, "Afghan War Is a Lab for U.S. Innovation," *Washington Post*, March 26, 2002.

19. Lambeth, *Air Power Against Terror*, 260.

20. Sean Naylor, *Not a Good Day to Die: The Untold Story of Operation Anaconda* (New York: Berkeley, 2005), 263.

21. Lambeth, *Air Power Against Terror*, 145.

22. Defense Science Board, *1996 Summer Study Task Force on Tactics and Technology for 21st Century Military Superiority* (Washington, DC: Department of Defense, 1996).

23. For a good overview of the exercise, see James A. Lasswell, "Assessing Hunter Warrior," *Armed Forces Journal International*, May 1997, 14–15.

24. John F. Schmitt, "A Critique of the Hunter Warrior Concept," *Marine Corps Gazette* 82, no. 6 (June 1998): 14–15.

25. Gary Anderson, "Infestation Tactics and Operational Maneuver from the Sea: Where Do We Go from Here?" *Marine Corps Gazette* 81, no. 9 (September 1997): 70.

26. Owen O. West, "Who Will Be the First to Fight?" *Marine Corps Gazette* 87, no. 5 (May 2003): 54.

27. John H. Hay Jr., *Vietnam Studies: Tactical and Materiel Innovations* (Washington, DC: Department of the Army, 1989), 52.

28. Thom Shanker, "The Edge of Night," *New York Times Magazine*, December 1, 2002.

29. Naylor, *Not a Good Day to Die*, 207.

30. Lambeth, *Air Power Against Terror*, 252.

31. Ross Kerber, "U.S. Bombs Seen Smarter, Cheaper," *Boston Globe*, October 3, 2003.

32. Eric E. Theisen, *Ground-Aided Precision Strike: Heavy Bomber Activity in Operation Enduring Freedom* (Maxwell AFB, AL: Air University Press, 2003), 1, 10.

33. Lambeth, *Air Power Against Terror*, 144.

34. Headquarters, United States Air Force, *Operation Anaconda: An Air Power Perspective* (Washington, DC: Department of Defense, 2005), 61.

35. Lambeth, *Air Power Against Terror*, 159.

36. 9/11 Report, 189–90.

37. Ibid., 211–12.

38. *Aerospace Daily and Defense Report*, October 13, 2004.

39. Naylor, *Not a Good Day to Die*, 357.

40. Schroen, *First In*, 166–67.

41. Lambeth, *Air Power Against Terror*, 278.

42. Douglas Jehl, "Remotely Controlled Craft Part of U.S.-Pakistan Drive Against Al Qaeda, Ex-Officials Say," *New York Times*, May 16, 2005.

43. Douglas Jehl and Mohammad Khan, "Top Qaeda Aide Is Called Target in U.S. Air Raid," *New York Times*, January 14, 2006.

44. Naylor, *Not a Good Day to Die*, 307.

45. Ibid., 152.

46. Lambeth, *Air Power Against Terror*, 326.

47. Tommy Franks, *American Soldier* (New York: HarperCollins, 2004), 290–96.

48. Ibid., 294.

49. Lambeth, *Air Power Against Terror*, 203.

50. Jeffrey Record, "Collapsed Countries, Casualty Dread, and the New American Way of War," *Parameters* (Summer 2002): 4–23.

51. Arthur K. Cebrowski and Thomas P. M. Barnett, "The American Way of War," *Transformation Trends*, January 13, 2003.

52. Quoted in *Military Transformation: A Strategic Approach* (Washington, DC: Department of Defense, 2003), 28.

53. "President Speaks on War Effort to Citadel Cadets," at www.whitehouse.gov/news/releases/2001/12/print/20011211–6.html (accessed December 11, 2001).

54. Ibid.

55. Stephen Biddle, *Afghanistan and the Future of Warfare: Implications for Army and Defense Policy* (Carlisle, PA: U.S. Army War College, 2002), vii–ix.

56. See, for example, Usama Bin Laden's 1998 *fatwa*, at www.fas.org/irp/world/para/docs/980223-fatwa.htm (accessed October 3, 2005).

57. Gregory Fontenot, E. J. Degan, and David Tohn, *On Point: The United States Army in Operation Iraqi Freedom* (Ft. Leavenworth, KS: Combat Studies Institute Press, 2004), 100.

58. Barry R. McCaffrey, "A Time to Fight," *Wall Street Journal*, April 1, 2003. Given subsequent events, it is worth noting that McCaffrey was discussing the requirement to defeat the Iraqi army, not to stabilize or occupy Iraq.

59. Thom Shanker and John Tierney, "Head of Military Denounces Critics of Iraq Campaign," *New York Times*, April 2, 2003.

60. Fontenot, Degan, and Tohn, *On Point*, 90.

61. Michael R. Gordon and Bernard E. Trainor, *Cobra II: The Inside Story of the Invasion and Occupation of Iraq* (New York: Pantheon Books, 2006), 177, 409.

62. Fontenot, Degan, and Tohn, *On Point*, 153.

63. Robinson, *Masters of Chaos*, chapters 9 and 13; Gordon and Trainor, *Cobra II*, chapter 17.

64. Fontenot, Degan, and Tohn, *On Point*, 89.

65. Dennis Cauchon, "Why U.S. Casualties Were Low," *USA Today*, April 21, 2003; Oscar Avila, "Allies Won with Few Casualties," *Chicago Tribune*, May 3, 2003.

66. John Keegan, "Saddam's Utter Collapse Shows This Has Not Been a Real War," *Daily Telegraph*, April 8, 2003.

67. *CENTAF Assessment and Analysis Division, Operation IRAQI FREEDOM—By the Numbers*, April 30, 2003, 3, 7.

68. Ibid., 11.

69. Kevin M. Woods et al., *Iraqi Perspectives Project: A View of Operation Iraqi Freedom from Saddam's Senior Leadership* (Norfolk, VA: U.S. Joint Forces Command, 2006), 125.

70. Although JDAM was able to operate through weather, the sandstorm still affected the ability of the United States to launch air strikes. U.S. Central Command cancelled one-quarter of planned air strikes on March 27, for example, because of weather. Gordon and Trainor, *Cobra II*, 324.

71. Fontenot, Degan, and Tohn, *On Point*, 255.

72. Bill Gertz, "Signal Jamming a Factor in Future Wars, General Says," *Washington Times,* July 16, 2004; Gordon and Trainor, *Cobra II,* 324.

73. Franks, *American Soldier,* 446.

74. Richard J. Dunn, *Blue Force Tracking: The Afghanistan and Iraq Experience and Its Implications for the U.S. Army* (Washington, DC: Northrop Grumman, 2004), 5.

75. Ibid., 7.

76. Ibid., 9, 11.

77. Adam Grissom, "The Future of Military Innovation Studies," *Journal of Strategic Studies* 29, no. 5 (October 2006): 929.

78. Fontenot, Degan, and Tohn, *On Point,* 97, 65.

79. Dennis M. Gormley, "Missile Defense Myopia: Lessons from the Iraq War," *Survival* 45, no. 4 (Winter 2003/2004): 61, 63, 66.

80. "We Got Nothing Until They Slammed Into Us," *Technology Review* (November 2004), 38.

81. Fontenot, Degan, and Tohn, *On Point,* 11.

82. Ibid., 174.

83. "We Got Nothing," 38; Greg Grant, "Network Centric Blind Spot," *Defense News,* September 12, 2005.

84. David Zucchino, *Thunder Run: Three Days in the Battle for Baghdad* (London: Atlantic Books, 2004), 12.

85. Ibid., 35.

86. Juan O. Tamayo, "Iraqis Seem Unaware of Enemy Location," *Miami Herald,* April 9, 2003.

87. Woods et al., *Iraqi Perspectives Project,* chapter 6.

88. "President Bush Outlines Progress in Operation Iraqi Freedom," April 16, 2003, at www.state.gov/p/nea/rls/rm/19709.htm (accessed April 2, 2006).

89. Max Boot, "The New American Way of War," *Foreign Affairs* 82, no. 4 (July/August 2003): 42–44.

90. Michael E. O'Hanlon and Jason H. Campbell, "Iraq Index: Tracking Variables of Reconstruction and Security in Post-Saddam Iraq," at www.brookings.edu/iraqindex (accessed September 21, 2007).

91. Brigadier Nigel Aylwin-Foster, "Changing the Army for Counterinsurgency Operations," *Military Review* (November/December 2005): 10.

92. Carol J. Williams, "Soldiers Get Extra Layer of Defense," *Los Angeles Times,* July 29, 2005.

93. David S. Cloud, "Insurgents Using Bigger, More Lethal Bombs, U.S. Officers Say," *New York Times,* August 4, 2005.

94. Steven Komarow, "Tanks Take a Beating in Iraq," *USA Today,* March 30, 2005.

95. John M. Donnelly, "New Countermeasure for Roadside Bombs Nearly Ready, House Chairman Says," *CQ Today,* March 11, 2005.

96. Megan Scully, "Hunting for Solutions to IEDs: U.S. Army Rounds Up Resources to Stop Roadside Bombs," *Defense News,* September 27, 2004.

97. David A. Fulghum, "Looking for the Silver Bullet," *Aviation Week and Space Technology,* May 9, 2005.

98. Ronald J. Glasser, "A War of Disabilities," *Harper's,* July 1, 2005.

99. Tom Infield, "The Stunning Success of Battlefield Medicine," *Philadelphia Inquirer,* December 11, 2005.

100. Glasser, "A War of Disabilities," 59.

Conclusion

This book has explored the interplay of technology and the culture of the U.S. armed services in the context of the strategic environment from 1945 to 2005. Its central argument has been that although the culture of the U.S. armed services both shaped and was shaped by technology, the services molded technology to suit their purposes more often than technology shaped them. This final chapter attempts to glean insights from this period for the contemporary debate over the prospect of major change in the conduct of warfare more broadly, and over the best course of action for the U.S. armed forces in dealing with such change more narrowly.

A Revolution in Warfare?

Over the last fifteen years, U.S. defense analysts have debated whether changes in technology, doctrine, and organization brought on by the information revolution presage a revolution in military affairs (RMA). This is but the most recent in a series of debates over the role of technology in the U.S. military. Current discussions echo those of the role of nuclear weapons in the U.S. armed forces after World War II, as well as those between the military reformers and defense traditionalists in the 1970s and 1980s.

On one side of the current debate are RMA enthusiasts who believe that we are witnessing a fundamental change in the character and conduct of war. Some argue that the exploitation of new technology will give the United States a significant military advantage over potential adversaries. As James R. Blaker, formerly a senior aide to Admiral William Owens, wrote in 1997, "The potency of the American RMA stems from new military systems that will create, through their interaction, in enormous military disparity between the United States and any opponent. Baldly stated, U.S. military forces will be able to apply military force with dramatically greater efficiency than an opponent, and do so with little risk to U.S. forces."[1] Others argue that the United States could find itself at a disadvantage at the hands of a future foe should it fail to transform its armed forces.[2]

On the other side are the RMA skeptics, who believe that faith in advanced technology is misguided, if not dangerous. Some argue that technology rarely delivers on its promises. Others feel that a focus on technology diverts attention from more important determinants of military effectiveness, such as training.[3]

Each pole in this debate offers a simplistic view of the role of technology in war in general and of its place in the U.S. military in particular. Enthusiasts, for their part, tend to overstate both the magnitude of change wrought by technology as well as the rate at which new technology can be assimilated into military organizations. As this book has shown, the culture of the armed services has tended to shape which technologies they have pursued as well as how they have employed new weapons on the battlefield. Moreover, the process of acquiring and breaking in weapon systems has often delayed their effective use on the battlefield. For example, although PGMs and UAVs saw widespread use in Vietnam, they really only came of age during the wars of the 1990s.

It is similarly true that the more breathless predictions of dramatic changes to the conduct of war wrought by technology have failed to materialize. Nuclear weapons and intercontinental ballistic missiles did not render conventional forces obsolete during the Cold War. Predictions of an "electronic battlefield" dominated by unmanned sensors and precision-guided munitions, first enunciated during the Vietnam War, are only now coming to fruition.

If the enthusiasts are guilty of hyping technology, the skeptics have all too often discounted the role of technology in war. Although technology is not the only—or necessarily the most important—determinant of success, its effects should not be ignored. Technology has played an important role in U.S. military success in each of the conflicts this book has examined. Technology gave the United States an edge over the Soviet Union in the

Cold War. Technology also increased the effectiveness of U.S. forces in Vietnam, though it could not salvage a flawed strategy. And over the past fifteen years technology has helped create a series of lopsided battlefield outcomes between the United States and Iraq (twice), Serbia (twice), and Afghanistan.

A Glass Half Full?

Whether one is a technology enthusiast or a skeptic colors one's view of the U.S. armed forces. From the enthusiast's perspective, the U.S. armed forces have experienced little change over the past six decades. To a large extent, today's armed forces resemble those that fought and won World War II. Manned aircraft dominate war in the air, and on land the main battle tank is the king of the battlefield. The aircraft carrier remains the capital ship. Indeed, one of the more noteworthy changes in the U.S. military in the last decade in a half has been the denuclearization of the U.S. military. To a remarkable extent, nuclear weapons, which were so central to the U.S. armed forces during the Cold War, have become marginal to the services that once embraced them so enthusiastically.

From another perspective, however, the U.S. armed forces have undergone drastic change. Today's manned aircraft, with reduced signatures to limit detection and armed with precision-guided munitions, are orders of magnitude more effective than their Vietnam-era predecessors armed with unguided weapons, let alone their World War II forebears. According to a 1993 Defense Science Board study, for example, for many target types, a ton of PGMs has replaced twelve to twenty tons of unguided munitions.[4] Whereas it usually took multiple sorties of aircraft dropping unguided weapons to destroy a target, today a single aircraft armed with PGMs can destroy several.

Today's tanks also are much more effective than those of previous generations, with the ability to see and shoot farther and more accurately than earlier models. They have also proven to be highly resistant to damage. They are also networked, giving their occupants a better understanding of where they, other friendly units, and enemy forces are located on the battlefield.

Although today's aircraft carriers represent incremental changes in design over those that fought in the Pacific campaign in World War II, the carrier strike groups of which they form the heart are far more effective, with air wings that have longer range and can deliver ordnance with precision, destroyers and cruisers whose phased-array radars give them the ability to detect objects at great distances (including in outer space), and land-attack cruise missiles that allow naval forces to strike far inland.

Perhaps nowhere is the impact of technology more apparent than in the infantry, the least technologically intensive branch of the U.S. armed forces.[5] The soldiers and marines who fought World War II were largely equipped with .303 caliber M-1 Garand semiautomatic rifles. The steel helmets they wore would protect them against shell fragments, but not from pistol or rifle bullets. To communicate, officers relied upon radiomen who lugged bulky radios.

The generation of infantrymen that fought in Vietnam in many ways resembled their World War II predecessors. To be sure, they were equipped with more effective weapons, primarily the M-16 automatic rifle. Squads possessed better yet still bulky handheld radios, and they could receive air support more reliably. In other ways, however, the Vietnam-era infantryman resembled his World War II counterpart. He still relied on a compass for navigation, for example, and had limited ability to operate at night.

Today's infantrymen, by contrast, is equipped with a helmet that can protect him against small-arms fire and body armor that can do the same. His M-4 automatic rifle, a direct descendant of the M-16, is equipped with optics that can make a mediocre shot into a marksman. He has night-vision goggles that allow him to see at night, a global positioning system (GPS) receiver that tells him where he is located on the battlefield, and a Blue Force Tracker that allows others to do likewise. His radio, which is now small enough to fit inside his helmet, allows him to communicate with his entire squad. This radio, together with laser designators and GPS receivers, allow him not only to mass the firepower of his unit, but also to bring down enormous destruction from artillery and aircraft.

What accounts for much of this change is the growth and spread of information technology. It is information technology, in the shape of precision guidance, that permits aircraft to strike targets with great accuracy. It is information technology, in the form of situational awareness systems such as Blue Force Tracker, that permits ground formations to coordinate their operations more effectively. It is information technology, in the form of networked communications, that allows carrier battle groups to operate dispersed and yet mass their firepower. And it is the pervasiveness of information technology, as well as the differing international and bureaucratic contexts, that illuminates the key differences between the information revolution and the nuclear revolution, the two periods of major change that mark the beginning and end of this study.

First, nuclear weapons and long-range delivery vehicles served as a tangible manifestation of the nuclear revolution. Indeed, the nuclear revolution represents the most clear-cut case of technology affecting the conduct of war in recent centuries. The engine of change for the information revolution, by contrast, is not a technology as concrete as a single weapon, but rather something as pervasive as information technology. Second, the strategic context of these developments differed. The nuclear revolution coincided with the beginning of the competition with the Soviet Union that made exploiting new ways of war an imperative. By contrast, the information revolution coincided with the end of the competition with the Soviet Union and questions over the shape of the future security environment. Third, the bureaucratic context differed. The nuclear revolution occurred during a period of significant change within the U.S. military, as roles and missions were up for competition. The information revolution, by contrast, is taking place in an era in which service roles and missions had largely been settled.

Evidence of change to the character of conflict falls into five categories. First, recent conflicts have demonstrated new ways of war. This is perhaps most apparent in the growing use of precision guided munitions (PGMs) since 1991: whereas 8 percent of the munitions employed during the Gulf War were guided, 29 percent of those used over Kosovo in eight years later, 60 percent of those used in Afghanistan ten years later, and 68 percent of those used in Iraq twelve years later were guided. As discussed in chapter 6, the widespread use of GPS-guided munitions such as the Joint Direct Attack Munition (JDAM) has been particularly noteworthy. In contrast with the laser-guided bombs that had been in use since Vietnam, such weapons allow aircraft to strike at night and through inclement weather.

Another sign of the changing character of war is the growing use of unmanned systems, both for reconnaissance and surveillance and, increasingly, strike missions. The U.S. military had only two operational types of UAVs in the year 2000, but at least twelve different systems are expected to be in active service by 2015.[6] As of August 2005, U.S. forces were operating approximately 1,500 unmanned aircraft in Iraq and Afghanistan.[7] This increase in UAV inventories is reflected in the Defense Department's annual budget request for these systems, which grew from $336 million in 2001 to $2.2 billion in 2005.[8] UAVs are also becoming more autonomous in their operations—increasingly responding to preprogrammed computer instructions entered through a keyboard, rather than direct remote control by pilots using a traditional "stick."

The use of UAVs for strike operations is on the rise as well. Delivery of lethal weapons from UAVs has been demonstrated in combat in Afghanistan and Iraq, and classes of UAVs are being developed specifically for weapons delivery—including an unmanned long-range bomber. The

2006 *Quadrennial Defense Review* (QDR) report stated that nearly half of the future long-range strike force will be unmanned.[9] Even the process of weapons delivery is likely to be increasingly automated. Boeing's unmanned X-45A, for instance, has already demonstrated the autonomous ability to make route selection to a general target location, and then identify and attack previously identified targets within an area as large as thirty miles by sixty miles.

A second trend involves the changing structure of military organizations. The availability of PGMs has allowed air forces to substitute increasingly for artillery. This has, in turn, changed the historical relationship between ground and air forces. In Kosovo, Afghanistan, and Iraq, ground forces served to fix enemy forces for engagement from the air. In some cases, infantry units have acted as sentient sensors, identifying and targeting enemy units. Although such an approach was most prominent in the "Afghan model" that was employed during Operation Enduring Freedom, it continues in a milder form in Iraq today.

There is perhaps no better example of the changing relationship between ground and air forces than the operation that killed the leader of Al Qaeda in Iraq, Abu Musab al-Zarqawi, in June 2006. Rather than storm the safehouse where al-Zarqawi was located, U.S. special operations forces reportedly identified the location and called in the air strike that destroyed the building.

Third, these changes challenge the identity of parts of the armed forces. Because GPS-guided weapons require much less operator involvement, they threaten to transform attack aircraft pilots into nothing more than glorified truck drivers. The widespread employment of UAVs and UCAVs presages an even more dramatic challenge to the identity of the pilot. Many of the UAVs operating over Afghanistan and Iraq are controlled not from the theater, but by operators located at Nellis Air Force Base outside of Las Vegas.

The current period contrasts in many ways with past periods of large-scale change. Past innovations have been marked by the rise of new specialties and organizations. The decades between the two world wars, for example, saw the emergence of aviators (both land-based and naval) and tankers. The nuclear revolution saw the appearance of missile operators and nuclear submariners. The current period has yet to yield similar new branches and career paths. Instead, it is leading to a redefinition of various existing elements of the military.

Fourth, changes to the character and conduct of war are reflected in the balance of power. Mastery of advanced technology has given the United States a substantial conventional advantage over the range of plausible adversaries. The experience of the past fifteen years contains ample evidence that the United States can defeat conventional militaries handily. Moreover, the United States possesses an absolute advantage in a number of areas. The U.S. advantage in anti-armor warfare, for example, is such that it is difficult to imagine an armored force that could threaten U.S. forces. It is also difficult to imagine a surface fleet that could compete with the U.S. Navy.

Adversaries have, of course, adopted countermeasures to America's conventional edge. Some have sought nuclear, biological, or chemical weapons in an effort to deter the United States or level the playing field should war come. Others have sought to battle the United States irregularly, adopting terrorism or guerrilla warfare strategies. Both remain significant challenges. Nonetheless, they should not obscure the magnitude of the U.S. conventional advantage.

A fifth clue to the changing character of warfare lies in the poor track record of experts over the past fifteen years. Predicting the course and outcome of future wars in a time of peace is difficult. Sir Michael Howard has likened the task to that of a sailor navigating by dead reckoning through a "fog of peace."[10] The past decade and a half, however, has featured numerous conflicts that have provided participant and observer alike valuable information on the character of contemporary conflict. As chapters 5 and 6 have shown, however, military experts have done a generally poor job of predicting the course or outcome of these conflicts. Although it may be that the quality of expertise in the military field is declining, a more compelling explanation is that the character of war is changing in some significant ways.

Technology and the U.S. Officer Corps

In the end, technology is developed and used by organizations, and the culture of those organizations has a great deal to do with which technologies are developed and how they are used. If we are in a period of significant change, then how well equipped are the U.S. armed forces to exploit it? Are U.S. officers technological enthusiasts or skeptics?

Some have asserted that the culture of the U.S. armed forces emphasizes technology over other, less tangible, determinants of battlefield success, such as training and leadership.[11] To them, the U.S. armed forces have embarked upon a quest for the Holy Grail of high technology while ignoring the persistence of friction on the modern battlefield.[12] Others have argued, with equal force, that the U.S. military is reluctant to embrace new ways of war, particularly those that threaten existing weapons, doctrine, and organizations. Rather than adapt to the information age, they see the services as perpetuating increasingly outmoded approaches to combat.[13]

Surveys of U.S. officers that James R. FitzSimonds and I conducted in 2000, 2002, and 2006 found that the U.S. armed forces were highly supportive of information-age ways of war, at least in the abstract.[14] For example, 85 percent of the officers surveyed in 2000 believed that forces employing information-age technology, doctrine, and organizations would enjoy a substantial edge over those that do not. Seventy-five percent felt that new ways of war would give the United States dominance over the full range of adversaries. Substantial majorities of officers in 2000, 2002, and 2006 predicted that information-age ways of war would make it easier to use force with decisive results and a reduced risk of U.S. casualties. Indeed, such views persisted in the face of the protracted war in Iraq.

U.S. officers also believed strongly in the growing importance of space and cyberspace. Seventy-six percent of officers surveyed in 2000 and 79 percent of those surveyed in 2002 felt that within the next twenty years conflicts would include combat operations in or from space. Eighty-five percent of those surveyed in 2000 and 74 percent of those surveyed in 2002 believed that within the same period, computer-network attack would become a central feature of military operations. A large number felt that we either are undergoing or may be undergoing "radical" change.

What exactly "radical" change is, however, is open to interpretation. Officers tended to equate transformation with marginal improvements to current weapons and doctrine rather than the development of fundamentally new capabilities. A majority believed that today's dominant systems—tanks, manned aircraft, and aircraft carriers—would be as important in twenty years as they are today. They also predicted that changes in technology, doctrine, and organization would have a limited impact on career paths in the military: Whereas a large percentage of officers in 2002 and 2006 believed that an individual in their branch or specialty would require very different skills in 2020, less than one in five believed their specialties would actually be rendered obsolete in that period. And the vast majority of officers were unwilling to reduce force structure or readiness to invest in new approaches to warfare. Similarly, only a small minority of officers supported the creation of a new service for space or cyberspace operations.

It is likely, then, that the U.S. armed services will favor advanced technology in general, but will be particularly bullish on those systems that comport with existing mission areas. Of course, such "evolutionary" changes can have revolutionary results. Precision guidance and stealth are two examples of technological developments that have not threatened existing missions or communities but have also had far-reaching consequences. The key, in many cases, will be the ability of civilian and military leaders to get the services to view these new ways of war as amenable to service culture. Major General James M. Gavin's argument in favor of the widespread adoption of helicopters in the Army is an excellent case in point: he made the case for radical change on the conservative grounds that it would restore the Army's ability to execute traditional cavalry missions.[15]

This book also shows that it is possible for the services to adopt new ways of war, including the development of intercontinental ballistic missiles, land-attack cruise missiles, and reconnaissance satellites. To do so, however, new technology, doctrine, and organizations must solve an existing or projected operational or strategic problem. Success also requires the support of high-level civilian or military leaders.

Some will doubtless argue that the central importance of irregular warfare will dampen the utility of advanced technology in coming years. Insurgency and counterinsurgency are not synonymous, however, with low technology. The insurgents in Iraq, for example, exploit information technology widely, from the cell phones and computers that they use to plan and coordinate their attacks to the triggers of their improvised explosive devices and the global news media that they use to spread their message. Hezbollah similarly used very sophisticated technology against Israel during its conflict in southern Lebanon in the summer of 2006.[16] Moreover, advanced technologies such as night-vision goggles, Blue Force Tracking, UAVs, and PGM have given U.S. forces a significant advantage in combating insurgents. To take just one example, to the extent that the rate and level of U.S. casualties influences public support for U.S. military operations abroad, the advent of modern body armor capable of protecting soldiers against assault rifle rounds, which has saved numerous lives in Afghanistan and Iraq, has had a strategic impact.

Of course, technology is only as effective as the strategy it serves. As the Cold War demonstrates, advanced technology can serve as an important element of a successful long-term strategy. By contrast, the Vietnam War demonstrates that even technological advantage cannot deliver victory when harnessed to a flawed strategy.

The United States will remain the most powerful state in the world for the foreseeable future. Its armed forces will need to wage a protracted war against jihadist extremists while also preparing for the possibility of a high-intensity conflict against a capable adversary. Indeed, balancing the very different capabilities required to confront near-term and far-term threats is one of the central challenges that U.S. defense planners face.

In some ways, the long war against jihadist extremists exists comfortably within the framework of the American way of war. In particular, it fits America's propensity for unlimited political objectives as well as its penchant for waging war against evil. Similarly, advanced technology may give the United States advantages against future foes, albeit in novel ways. Throughout much of history, for example, operations at night and in forbidding climate and terrain have favored the weaker side. Both against Iraq in 1991 and 2003 and in Afghanistan in 2001, their technological advantage allowed U.S. forces to operate in ways that their adversaries did not expect. During the Gulf War, for example, use of GPS for precision navigation allowed the U.S. military units to traverse trackless desert while Iraqi forces remained largely confined to roads. In that war and again in the campaign against the Taliban and Al Qaeda in Afghanistan, the U.S. advantage in night vision allowed it to operate freely at night when its adversaries could not.

In other ways, however, success in the current conflict will likely demand modes of operation that differ significantly form America's strategic traditions. First, it is doubtful that a direct approach will work in the future. Overthrowing governments that harbor terrorists is likely to be a favored option only in a small number of cases. Instead, the United States will have to use cooperation with local officials, law-enforcement methods, and covert operations to root out terrorists. Second, America's traditional reliance on firepower-intensive strategies may prove counterproductive in a conflict in which maintaining some level of popular support is necessary. Finally, the U.S. military needs to strengthen other areas of competency, such as those associated with special operations and stability operations. Over time, these changes may considerably alter our notion of what the American way of war is.

Notes

1. James R. Blaker, "The American RMA Force: An Alternative to the QDR," *Strategic Review* (Summer 1997): 22.
2. Office of the Secretary of Defense, *Transformation Planning Guidance* (Washington, DC: Department of Defense, 2003), 4–5.
3. Stephen Biddle, "Victory Misunderstood: What the Gulf War Tells Us About the Future of Conflict," *International Security* 21, no. 2 (Fall 1996): 139–79.
4. Benjamin S. Lambeth, *The Transformation of American Air Power* (Ithaca, NY: Cornell University Press, 2000), 160.
5. Eliot A. Cohen makes much the same point in "Change and Transformation in Military Affairs," *Journal of Strategic Studies* 27, no. 3 (September 2004): 403.
6. Office of the Secretary of Defense, *Unmanned Aircraft Systems Roadmap: 2005–2030* (Washington, DC: OSD, 2005), 3.

7. United States Government Accountability Office, *Unmanned Aircraft Systems: DoD Needs to More Effectively Promote Interoperability and Improve Performance Assessments*, GAO-06-49 (December 2005), 7.

8. Ibid., 7.

9. Donald Rumsfeld, *Quadrennial Defense Review Report* (Washington, DC: OSD, 2006), 46.

10. Michael Howard, "Military Science in an Age of Peace," *Journal of the United Services Institute for Defense Studies* 119, no. 1 (March 1974): 4.

11. Warren Caldwell, "Promises, Promises," *Proceedings* (January 1996).

12. Mackubin T. Owens, "Technology, the RMA, and Future War," *Strategic Review* (Spring 1998): 63–70.

13. Eliot A. Cohen, "Defending America in the Twenty-first Century," *Foreign Affairs* 79, no. 6 (November/December 2000): 40–56; Andrew F. Krepinevich, "Why No Transformation?" *Joint Force Quarterly* (Autumn/Winter 1999–2000): 97–101.

14. The detailed findings of the 2000 survey were published in Thomas G. Mahnken and James R. FitzSimonds, *The Limits of Transformation: Officer Attitudes Toward the Revolution in Military Affairs* (Newport, RI: Naval War College Press, 2003). Detailed findings of Army officer attitudes from the 2002 survey were published in Thomas G. Mahnken and James R. FitzSimonds, "Tread-Heads or Technophiles? Army Officer Attitudes Toward Transformation," *Parameters* (Summer 2004): 57–72.

15. James M. Gavin, "Cavalry, and I Don't Mean Horses," *Harper's Magazine*, April 1954.

16. Edward Cody and Molly Moore, "'The Best Guerrilla Force in the World': Analysts Attribute Hezbollah's Resilience to Zeal, Secrecy and Iranian Funding," *Washington Post*, August 14, 2006.

Printed in the USA
CPSIA information can be obtained
at www.ICGtesting.com
JSHW011758080824
67816JS00022B/466